C0-AUR-516

THE MINNESOTA LECTURES ON THE STRUCTURE AND DYNAMICS OF THE MILKY WAY

A SERIES OF BOOKS ON RECENT DEVELOPMENTS IN ASTRONOMY AND ASTROPHYSICS

Printed by BookCrafters, Inc.

First published 1993

Library of Congress Catalog Card Number: 92-76080
ISBN 0-937707-58-9

D. Harold McNamara, Managing Editor of Conference Series
408 ESC Brigham Young University
Provo, UT 84602
801-378-2298

A SERIES OF BOOKS ON RECENT DEVELOPMENTS IN ASTRONOMY AND ASTROPHYSICS

Vol. 1-Progress and Opportunities in Southern Hemisphere Optical Astronomy: The CTIO 25th Anniversary Symposium
ed. V. M. Blanco and M. M. Phillips ISBN 0-937707-18-X

Vol. 2-Proceedings of a Workshop on Optical Surveys for Quasars
ed. P. S. Osmer, A. C. Porter, R. F. Green, and C. B. Foltz ISBN 0-937707-19-8

Vol. 3-Fiber Optics in Astronomy
ed. S. C. Barden ISBN 0-937707-20-1

Vol. 4-The Extragalactic Distance Scale: Proceedings of the ASP 100th Anniversary Symposium
ed. S. van den Bergh and C. J. Pritchet ISBN 0-937707-21-X

Vol. 5-The Minnesota Lectures on Clusters of Galaxies and Large-Scale Structure
ed. J. M. Dickey ISBN 0-937707-22-8

Vol. 6-Synthesis Imaging in Radio Astronomy: A Collection of Lectures from the Third NRAO Synthesis Imaging Summer School
ed. R. A. Perley, F. R. Schwab, and A. H. Bridle ISBN 0-937707-23-6

Vol. 7-Properties of Hot Luminous Stars: Boulder-Munich Workshop
ed. C. D. Garmany ISBN 0-937707-24-4

Vol. 8-CCDs in Astronomy
ed. G. H. Jacoby ISBN 0-937707-25-2

Vol. 9-Cool Stars, Stellar Systems, and the Sun. Sixth Cambridge Workshop
ed. G. Wallerstein ISBN 0-937707-27-9

Vol. 10-The Evolution of the Universe of Galaxies. The Edwin Hubble Centennial Symposium
ed. R. G. Kron ISBN 0-937707-28-7

Vol. 11-Confrontation Between Stellar Pulsation and Evolution
ed. C. Cacciari and G. Clementini ISBN 0-937707-30-9

Vol. 12-The Evolution of the Interstellar Medium
ed. L. Blitz ISBN 0-937707-31-7

Vol. 13-The Formation and Evolution of Star Clusters
ed. K. Janes ISBN 0-937707-32-5

Vol. 14-Astrophysics with Infrared Arrays
ed. R. Elston ISBN 0-937707-33-3

Vol. 15-Large-Scale Structures and Peculiar Motions in the Universe
ed. D. W. Latham and L. A. N. da Costa ISBN 0-937707-34-1

Vol. 16-Atoms, Ions and Molecules: New Results in Spectral Line Astrophysics
ed. A. D. Haschick and P. T. P. Ho ISBN 0-937707-35-X

Vol. 17-Light Pollution, Radio Interference, and Space Debris
ed. D. L. Crawford ISBN 0-937707-36-8

Vol. 18-The Interpretation of Modern Synthesis Observations of Spiral Galaxies
ed. M. Duric and P. C. Crane ISBN 0-937707-37-6

Vol. 19-Radio Interferometry: Theory, Techniques, and Applications, IAU Colloquium 131
ed. T. J. Cornwell and R. A. Perley ISBN 0-937707-38-4

Vol. 20-Frontiers of Stellar Evolution, celebrating the 50th Anniversary of McDonald Observatory
ed. D. L. Lambert ISBN 0-937707-39-2

Vol. 21-The Space Distribution of Quasars
ed. D. Crampton ISBN 0-937707-40-6

Vol. 22-Nonisotropic and Variable Outflows from Stars
ed. L. Drissen, C. Leitherer, and A. Nota — ISBN 0-937707-41-4

Vol. 23-Astronomical CCD Observing and Reduction Techniques
ed. S. B. Howell — ISBN 0-937707-42-4

Vol. 24-Cosmology and Large-Scale Structure in the Universe
ed. R. R. de Carvalho — ISBN 0-937707-43-0

Vol. 25-Astronomical Data Analysis Software and Systems I
ed. D. M. Worrall, C. Biemesderfer, and J. Barnes — ISBN 0-937707-44-9

Vol. 26-Cool Stars, Stellar Systems, and the Sun, Seventh Cambridge Workshop
ed. M. S. Giampapa and J. A. Bookbinder — ISBN 0-937707-45-7

Vol. 27-The Solar Cycle
ed. K. L. Harvey — ISBN 0-937707-46-5

Vol. 28-Automated Telescopes for Photometry and Imaging
ed. S. J. Adelman, R. J. Dukes, Jr., and C. J. Adelman — ISBN 0-937707-47-3

Vol. 29-Workshop on Cataclysmic Variable Stars
ed. N. Vogt — ISBN 0-937707-48-1

Vol. 30-Variable Stars and Galaxies, in honor of M. S. Feast on his retirement
ed. B. Warner — ISBN 0-937707-49-X

Vol. 31-Relationships Between Active Galactic Nuclei and Starburst Galaxies
ed. A. V. Filippenko — ISBN 0-937707-50-3

Vol. 32-Complementary Approaches to Double and Multiple Star Research, IAU Colloquim 135
ed. H. A. McAlister and W. I. Hartkopf — ISBN 0-937707-51-1

Vol. 33-Research Amateur Astronomy
ed. S. J. Edberg — ISBN 0-937707-52-X

Vol. 34-Robotic Telescopes in the 1990s
ed. A. V. Filippenko — ISBN 0-937707-53-8

Vol. 35-Massive Stars: Their Lives in the Interstellar Medium
ed. J. P. Cassinelli and E. B. Churchwell — ISBN 0-937707-54-6

Vol. 36-Planets and Pulsars
ed. J. A. Phillips, S. E. Thorsett, and S. R. Kulkarni — ISBN 0-937707-55-4

Vol. 37-Fiber Optics in Astronomy II
ed. P. M. Gray — ISBN 0-937707-56-2

Vol. 38-New Frontiers in Binary Star Research
ed. J. C. Leung and I. S. Nha — ISBN 0-937707-57-0

Vol. 39-The Minnesota Lectures on the Structure and Dynamics of the Milky Way
ed. Roberta M. Humphreys — ISBN 0-937707-58-9

Inquiries concerning these volumes should be directed to the:

Astronomical Society of the Pacific
CONFERENCE SERIES
390 Ashton Avenue
San Francisco, CA 94112-1722
415-337-1100

ASTRONOMICAL SOCIETY OF THE PACIFIC
CONFERENCE SERIES

Volume 39

THE MINNESOTA LECTURES ON THE STRUCTURE AND DYNAMICS OF THE MILKY WAY

Edited by
Roberta M. Humphreys

Table of Contents

THE MINNESOTA LECTURES ON THE STRUCTURE AND DYNAMICS OF THE MILKY WAY

This volume is the proceedings of a lecture series in the Astronomy Department at the University of Minnesota during the spring of 1992 on the structure and dynamics of our galaxy. The proceedings of two similar courses have been published: a course on active galactic nuclei in Publications of the Astronomical Society of the Pacific **98**, p. 129 (1986) and a course on Cluster of Galaxies and Large Scale Structure as volume 5 in the Astronomical Society of the Pacific Conference Series (1988).

The lecture series was offered as an advanced course at the graduate level and was attended by both graduate students and faculty. The review papers in this volume are based on the lectures which were presented as two 90 minute talks. Since this was a graduate course problem sets and a final exam were given. A list of questions and problems is included at the end of this volume.

I am grateful to the speakers for participating in the lecture series. Their visit to our department included informal meetings with the students, lunches and dinners as well as the formal lectures. We all appreciate their willingness to make their lecture material available through this volume.

Roberta M. Humphreys
October 1992

The Minnesota Lectures on Clusters of Galaxies and Large-Scale Structure
ASP Conference Series, Vol. 39, 1993
Roberta M Humphreys (ed.)

INTRODUCTION: HISTORICAL OVERVIEW AND THE CONCEPT OF STELLAR POPULATIONS

Roberta M. Humphreys
University of Minnesota, Department of Astronomy,
116 Church Street SE, Minneapolis, Minnesota

Much of the information in this introductory chapter can be found in many references. A primary one is Galactic Astronomy by Mihalas and Binney. This discussion is included here for completeness and to make this volume more useful as a general reference and text for a graduate seminar in galactic structure.

At the beginning of this century the "Kapteyn Universe" (Kapteyn and von Rhijn 1920, Kapteyn 1922) was the accepted model for the structure of our galaxy which was also considered by many at that time to be equivalent to the known universe. Kapteyn used the counts of stars versus magnitude, with proper motions for distance (r) estimates of some of the stars, and then simply assumed that the luminosity is proportional to $1/r^2$ ($\ell < 1/r^2$). The result was that the galaxy was found to be a flattened spheroid about five times longer in the plane than in the perpendicular direction with the sun slightly out of the plane at only 650 parsecs from the center. The star density was modelled to drop to 1% at 8500 pc in the plane and 1700 pc in the perpendicular. This was a very heliocentric model for the galaxy and did not include the effects of interstellar extinction.

In his book The Distribution of the Stars in Space, Bart Bok (1937) describes the history of star counts as a means to determining the geometric structure of the galaxy. In the classical work such as that by Kapteyn, astronomers tried to invert the fundamental equation of stellar statistics to determine the density distribution of the stars.

Star Counts refers to the fundamental statistical method used to construct models of the distribution of stars in our galaxy. In the method of star counts the number of stars in a given area as a function of magnitude is used to find their space density. The definitions and notation used here is common to many references:

A(m,b) is the number of stars at apparent magnitude m per unit magnitude per square degree at galactic latitude b. A(m,b) is simply determined by counting the number of stars in the range m - 1/2 to m + 1/2. A(m,b) dm is then the number of stars in the range (m, m+dm).

From the differential star counts A(m,b) we can also determine the integrated counts. N(m,b) is the cumulative number of stars per square degree with magnitudes less than or equal to m.

$$\frac{dN(m,b)}{dm} \equiv A(m,b) \tag{1}$$

or

$$N(m,b) = \int_{-\infty}^{m} A(m,b)dm \tag{2}$$

and

$$N(m,b) = \sum_{m=-2}^{m} A(m',b)$$

where N(m,b) gives the total number down to m + 1/2.

Kapteyn found a very slow growth in N(m) with distance from the Sun which led to his heliocentric model. If interstellar absorption is present then the distances to the stars will be overestimated producing an apparent drop in the stellar density from the Sun.

Not long after Kapteyn's model was proposed Harlow Shapley completed his historic study of the globular clusters in our galaxy that challenged all prior models of the galaxy. Shapley determined the distribution of globular clusters using distances from RR Lyrae stars. Although they wre distributed uniformly above and below the galactic plane, the globulars were concentrated in the direction of the constellation Sagittarius - the galactic bulge. This assymmetric distribution showed that the center of the galaxy was not near the Sun as in the Kapteyn universe.

In 1926 Lindblad (Lindblad 1927) presented a mathematical model of galactic rotation that confirmed Shapely's model of the galaxy by showing that the gravitational field in the Kapteyn model was too weak to retain the globular clusters. He proposed that the globulars were nearly at rest with respect to galactic rotation while the Sun and disk stars travel in circular orbits about the center. Thus the stars that we call `high velocity' appear to lag behind the Sun in a direction perpendicular to the galactic center. This is because the Sun belongs to a rapidly rotating system moving on a circular orbit while these `high velocity' stars belong to a slower rotating system.

During 1927-28 Jan Oort (Oort 1927, 1928) developed the equations of differential rotation and the expressions for Oort's constants. Bart Bok often told his students this anecdote: While Bok was a graduate student taking Oort's galactic structure course at the Leiden Observatory, Oort was working his way through Lindblad's paper. He apparently encountered some difficulty and asked the class for some time. When he returned he wrote down the now famous equations of differential rotation.

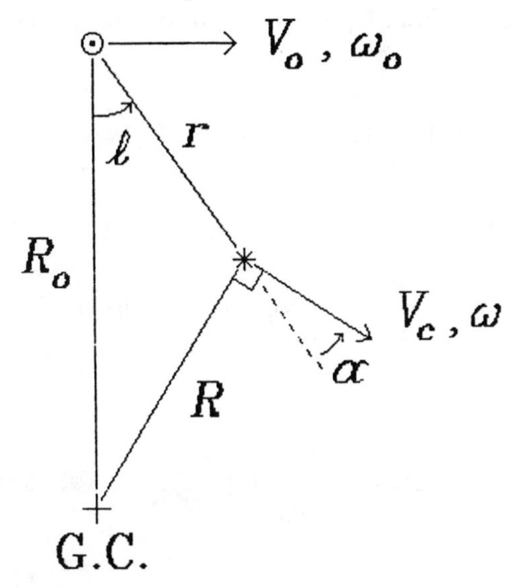

Figure 1. The basic figure for the different rotation of the galaxy.

The velocity due to galactic rotation at distance R from the galactic center is designated as V_c and the angular velocity of rotation is ω.

Then from Figure 1

$$V_c = R_o[\omega(R)-\omega(R_o)] \sin \ell \cos b \tag{3}$$

and the radial velocity along the line of sight is

$$v_r = V_c \cos \alpha - V_\odot \sin \ell \tag{4}$$

The detailed development of the equations of differential rotation and the derivation of Oort's constants can be found in many references. So here I just simply give the principal equations.

Using the law of sines equation (2) can be written as:

$$v_r = (\omega - \omega_o) R_o \sin \ell \qquad (5)$$

and if r, the distance from the Sun is small, then

$$v_r = (\omega - \omega_o) R \sin \ell \qquad (6)$$

Expanding ω-ω_o about R, equation (6) can be written

$$v_r = [(\frac{V_c}{dR})_{R_o} - \frac{V_o}{R_o}] (R-R_o) \sin \ell \qquad (7)$$

and for $r << R_o$, $R_o - R \simeq r \cos \ell$. Then:

$$v_r \simeq [\frac{V_o}{R_o} - (\frac{V_c}{dR})_{R_o}] r \sin \ell \cos \ell \qquad (8)$$

and defining:

$$A = \frac{1}{2} [\frac{V_o}{R_o} - (\frac{V_c}{dR})_{R_o}] \qquad \text{(9) Oort's constant A}$$

$$v_r = A_r \sin 2 \ell \qquad (10)$$

Thus because of differential rotation, the radial velocities of stars in the disk show a double sine wave.

Similarly defining

$$B = -\frac{1}{2} [\frac{V_o}{R_o} + (\frac{V_c}{dR})_{R_o}] \qquad \text{(11) Oort's Constant B}$$

$$\omega_o = \frac{V_o}{R_o} = A - B \qquad (12)$$

and

$$\left(\frac{V_c}{dR}\right)_{R_o} = -(A+B) \tag{13}$$

It wasn't until 1930 that Trumpler proved the existence of the extinction of starlight by interstellar dust even though it had been suspected for some time. Trumpler compared the distances of clusters of stars determined from their angular diameter with the distances inferred from their color-magnitude diagrams. He found that the color-magnitude diagrams systematically implied a larger distance and concluded that the stellar magnitudes had been dimmed by dust. Interstellar extinction thus explained the anomalous results from star counts in the Kapteyn model.

One of the most important developments in stellar and galactic astronomy occurred in 1944 when Walter Baade resolved the bulges of M31, M32, NGC205 plus NGC147 and NGC185 using the 100-inch telescope on Mt. Wilson. Baade's observations showed that the galactic bulges were dominated by a population of older red giant stars while the light of the spiral arms was mostly from luminous hotter stars. These two populations of stars had very different color-magnitude diagrams. On this basis Baade proposed the two stellar populations; I - the disk and spiral arms and II - the bulge and halo.

In addition to the differences in their color-magniutude diagrams the two basic populations have other distinguishing properties:

	Population I	Population II
Location:	disk and spiral arm	bulge, halo and globular clusters
Age:	very young to old	very old
Composition:	solar abundances	metal poor
Kinematics:	circular orbits, low velocity with respect to the sun	highly elliptical orbits, high velocity with respect to the sun

The early work resulted in the famous Vatican Conference (1957) on Stellar Populations in which five subgroups were distinguished by their different characteristics:

	Distance above plane (parsecs)	Vel. Disp. vert. dir. km/sec	Heavy Element abundances	Age (10^9 yrs)	Distribution
Halo Population II	2000	75	.003	≥6	Smooth
Intermediate Population II	700	25	.01	5-6	Smooth
Disk Population	400	17	.02	1.5-5	Smooth:
Order Population I	160	10	.03	0.1-1.5	Patchy
Extreme Population I	120	8	.04	<0.1	Very patchy, spiral arms

The concept of stellar populations has played a critical role in the development of modern studies of stellar evolution, galactic structure and galactic evolution. The articles in this volume by Don Terndrup and Ken Janes present a modern discussion of the stellar populations in our galaxy.

Perhaps the most convincing evidence that our galaxy must be a spiral galaxy comes from its flattened disk plus the wide-angle photographs that reveal its central bulge with a belt of obscurring material. These are of course the characteristics of spiral galaxies when viewed edge-on. From observations of other spirals we know that the youngest and most luminous stars are the best optical tracers of the spiral arms. W.W. Morgan and his collaborators (Morgan et al. 1952, 1953) used this fact when they mapped the distribution of the young associations and clusters and showed for the first time the presence of three spiral features in our region of the Galaxy. Morgan has said that one of his motivations for developing the famous MK system of two-dimensional spectral classification was to study the structure of the galaxy.

The young stars within a few kiloparsecs of the Sun are distributed in spiral features but optical mapping of spiral structure to much greater distances is severely limited by interstellar dust. The apparent solution came with the discovery of the 21cm of neutral hydrogen. In 1944 H.C. van de Hulst predicted the 21cm line of neutral hydrogen which was detected for the first time in 1951 by Ewan and Purcell. It wasn't until 1957 that Schmidt (1957) and Westerhout (1957) produced the first picture of the distribution of HI for the Northern sky and Kerr (1959) later published the first map of the Southern Sky. These analyses assumed optically thin gas and pure circular rotation. The results showed the HI distributed in fragmented, ribbon-like, nearly circular features about the galactic center.

The distances to the HI gas are "kinematic;" based on an assumed rotation curve for the galaxy and depend on a model for galactic rotation which then assumed pure circular motion. However non-circular motions of ~ ±10 km/sec are known to exist in both the stars and gas (Humphreys 1970, Burton 1976). These relatively small deviations lead to significant errors in the distances. In his paper in this volume, John Dickey presents the current work and status of the structure of the galaxy as revealed by the distribution of the gas including HI and CO.

Renewed interest in both the theory and observations of the spiral structure of our galaxy was stimulated by the development of the density wave theory (Lin et al. 1969). It provided a mathematical model with specific predictions that could be tested against observations. Today two basic types of spiral arms are recognized: the gravity or density wave arms and less regular flocculent arms produced by stochastic star formation. Ivan King reviews galactic dynamics and current outstanding questions.

During the early 1980's the entire subject of galactic structure went through a rennaissance ranging from a return to star count methods using modern models combined with the large datasets available from automated measuring machines to the detailed study of stellar populations in other galaxies now possible with modern optical and infrared detectors on large telescopes. The paper by Gerry Gilmore gives a current discussion of these methods and problems. Also during this same time, infrared and radio high resolution observations of the galactic center revealed not only its structure but also clues to the kinds of stars that are present. John Lacy's article summarizes the current status of observations of the galactic nucleus.

Thus the study of the structure of our galaxy once thought to be understood and a rather "statistical" subject is now a very important component of modern astrophysical efforts to understand the structure and evolution of galaxies.

REFERENCES

Baade, W. 1944, Ap.J., 100, 137.

Baade, W. 1944, Ap.J., 100, 147.

Bok, B.J. 1937, The Distribution of the Stars in Space (Chicago: University Chicago Press).

Burton, W.B. 1972, A. & A., 19, 51.

Humphreys, R.M. 1970, A.J., 75, 602.

Kapteyn, J.C. 1922, Ap.J., 55, 302.
Kapteyn, J.C. and van Rhijn, P.J. 1920, Ap.J., 52, 23.
Kerr, F.J., Hindman, J.V. and Gunn, C.S. 1959, Austr. J. Phys., 12, 1270.
Lin, C.C., Yuan, C. and Shu, F.H. 1969, Ap.J., 155, 721.
Lindblad, B. 1927, Mon. Not. Roy. Astron. Soc., 87, 553.
Morgan, W.W., Sharpless, S. and Osterbrock, D.E. 1952, A.J., 57, 3.
Morgan, W.W., Whitford, A.E. and code, A.D. 1953, Ap.J., 18, 318.
Oort, J.H. 1927, Bull. Astron. Inst. Netherlands, 3, 275.
Oort, J.H. 1928, Bull. Astron. Inst. Netherlands, 4, 269.
Schmidt, M. 1957, Bull. Astron. Inst. Netherlands, 13, 247.
Trumpler, R.J. 1930, Lick Obs. Bull., 14, 154.
Westerhout, G. 1957, Bull. Astron. Inst. Netherlands, 13, 201.

The Minnesota Lectures on Clusters of Galaxies and Large-Scale Structure
ASP Conference Series, Vol. 39, 1993
Roberta M Humphreys (ed.)

STELLAR POPULATIONS: BULGE, DISK, AND HALO

DONALD M. TERNDRUP[1]
Department of Astronomy, The Ohio State University, 174 W. 18th Ave., Columbus, OH 43210

INTRODUCTION

The concept of stellar populations has proven to be one of the most useful ideas of modern astronomy. Most simply, it is the idea that we can define a *population* as a set of stars that possess shared characterists such as composition, age, or kinematics, and that we can use the properties of the various stellar populations to determine the structure and evolutionary history of the Galaxy. Defined this way, we can see immediately that the characteristics we use to define a population should be independent of the evolution of individual stars, and instead be signatures of a common origin.

I would like to start with a brief historical overview, discuss the principal populations in the Galaxy, then conclude with some simple comparisons of these main populations. Excellent reviews of the subject have been written by King (1971), van den Bergh (1975), Mould (1982), Sandage (1986) and others. Sandage's article includes an especially thorough treatment of the original development and subsequent growth of the population concept, and I have drawn much of the following from his discussion.

The first modern definition of stellar populations was by Baade (1944), who defined two archetype populations – the now famous Populations I and II – based on color-magnitude diagrams and on the brightness of resolved stars in M31 and M32. In his original scheme, Population I is composed of stars found in the disk of our Galaxy and in the Magellanic Clouds; a signature of this population is the presence of highly luminous O- and B-type stars and open clusters. Population II is made up of stars with color-magnitude diagrams like galactic globular clusters, which lack luminous blue stars; another signature of Population II is the presence of short-period Cepheid variables. Throughout the next decade or so, it became clear that Population II was old and poor in *metals*[2], while Population I was metal rich and contains both young and old stars.

The key to Baade's argument was not simply that there were different types of stars; indeed, he stated in his 1944 paper that the two populations had been recognized in our Galaxy by Oort as early as 1926. The key was his determination of the absolute magnitudes of the brightest red giants in the

[1] Presidential Young Investigator.

[2] A metal is any element except hydrogen and helium. Throughout this chapter, *metal rich* will refer to stars with metal content like that of the Sun, while *metal-poor* will mean stars with a metal content a few percent or less than the Sun's.

bulge of M31 and in its companion galaxies M32 and NGC205. He derived absolute magnitudes near $M_{pg} = -1$ (photographic scale) for the brightest stars in all three galaxies, close to the value for the brightest stars in galactic globular clusters. Baade therefore included under Population II the stars in globular clusters, elliptical galaxies, and the bulges of our Galaxy and M31.

Baade then argued that the Hubble sequence of galaxy types E – Sa – Sb – Sc – Irr was simply a reflection of varying portions of Populations I and II. According to Baade, ellipticals were composed entirely of Population II, and irregulars entirely of Population I. Galaxies of intermediate type, such as our own (between Sb and Sc), had mixtures of populations, with Population I in the disk and Population II in the halo and bulge.

Baade (1951) later presented what he considered compelling evidence favoring the unification of bulge and globular cluster stars under Population II with his photometry of the RR Lyrae stars in the bulge of the Galaxy. Now the bulge is generally obscured from study at optical wavelengths because of large, patchy extinction from dust. There are, however, a few low-absorption "windows" through which bulge stars can be observed directly. In an especially clean window at the low galactic latitude of $b = -4°$, Baade showed that there were many RR Lyraes in the bulge, *as in the globular clusters.* The bulge RR Lyraes had a narrow distribution in magnitude, which demonstrated that the line of sight passed right through the bulge; the RR Lyrae density distribution had to be round like the bulges of other galaxies, but not flat like their disks. This window at $b = -4°$ has subsequently been the site of many investigations of the bulge, and has been named "Baade's Window" to honor his pioneering work.

Throughout the 1950's, as more data became available for star clusters and field stars, it was evident that the Population I/II concept was not a complete description of all stars in the Galaxy. For example, observations of the globular cluster system indicated that the classical Population II was not chemically homogeneous. It had been known for some time that the integrated light of the globular clusters varied in line strength, ranging from weak-lined, somewhat bluer light like that of F stars (very metal-poor clusters), to stronger-lined, redder light like G stars (moderately metal-poor clusters). The two groups had different spatial and kinematical distributions: the G-type clusters were concentrated more toward the galactic center or plane while the F-type clusters were more widely distributed. In addition, some stars that should have been in the halo from their orbits had color-magnitude diagrams similar to Population I disk objects like M67, rather than like those of globular clusters. Finally, spectroscopy of the central regions of M31 (summarized in Morgan and Osterbrock 1969) showed that absorption lines in the integrated light were strong; M31's bulge stars were therefore metal-rich. This was contrary to Baade's main point that the stars in Population II were all like globular cluster stars, and suggested that dividing the Galaxy's stars into merely two groups was too simple a scheme.

The various doubts and worries about the Population I/II scheme came to a head in the famous Vatican Conference of 1957 (see Blaauw 1965 for a discussion), during which Baade's original two Populations were replaced with five. The new scheme recognized that the original two-category plan did not sufficiently account for the range of *age* and *metal abundance* in either group, and also took into account the growing realization that a *kinematical* description of a population was of equal importance to characterizations by age and chemical composition.

Since the Vatican Conference, the field of stellar populations has been very active, and we now have a wealth of detailed information on populations throughout the Galaxy. Two important developments have occurred since that time. The first is the study of the relative abundances of the chemical elements; as I will discuss below, an understanding of abundance patterns can give us a variety of "clocks" to measure the rate of the production of the elements in the Galaxy. The second development is that we now have a good picture of the population in the galactic bulge, giving us a third major component of the galaxy to study; the bulge is now understood to be old and extremely metal-rich, and not metal-poor like the globular cluster system[3].

The important contribution from the Vatican Conference was not the increase in the number of identified populations in the Galaxy from two to five, but the realization that the boundaries of each group are not well defined. In other words, the characteristics of a given group of stars, depending on how we select them for study, can vary *continuously* and overlap the borders of our defined populations. By extension, then, one can view the study of stellar populations as the measurement of the distribution of stellar properties in a parameter space with axes of age, composition, spatial distribution, kinematics, etc.; these distributions may or may not be completely separable. That we can identify stellar populations at all probably indicates that the Galaxy used a relatively small number of physical processes in producing its stars and settling down to its current structure.

AN ASIDE: THE BIG PICTURE

One point worth making right away is that we have a considerably different picture of stellar populations in our Galaxy, where we have detailed information on spatial and orbital distributions, ages, metal abundances, etc., than we do for other galaxies, where we see only the integrated light distribution. In other galaxies, the terms bulge, disk, and halo refer to the light distribution; in the Milky Way these terms refer to particular groups of stars selected either by their spatial distribution, velocities, or some other classification. In this section, I will first discuss a common method for measuring the sizes of bulges and disks in other galaxies, then present a basic result from stellar kinematics that we will need later.

---

[3] As Frogel (1988) has discussed, one of the consequences of the metal-rich nature of the bulge is that the brightest stars on Baade's red photographic plates of M31 are up to four magnitudes less luminous than the most luminous stars bolometrically. In the more metal-poor globulars, the brightest red stars *are* the most luminous ones. That the two groups have nearly the same absolute red magnitude is coincidental, and was the source of Baade's error in merging bulge and globular cluster stars under one category.

The structure of spiral galaxies

The light distribution of external spiral galaxies can generally be described with reasonable precision as the sum of a flattened *disk* and a rounder *bulge*[4]. The disks of many spirals are well described by an exponential

$$\Sigma(R) = \Sigma_0 \exp(-R/R_d), \tag{1}$$

where R is radius measured along the disk from the center, R_d is the scale length of the disk, and Σ is the surface brightness measured in magnitudes arcsec^{-2}. We can then define the disk population of a galaxy as that part of the light distribution that is described by equation (1).

Ellipticals and the bulges of many spirals can be described by the de Vaucouleurs (1948) law

$$\begin{aligned} \Sigma(R) &= \Sigma_e 10^{\{-3.33[(R/R_e)^{1/4}-1]\}} \\ &= \Sigma_e \exp\{-7.67[(R/R_e)^{1/4} - 1]\}, \end{aligned} \tag{2}$$

where R_e is termed the *effective radius*; at $R = R_e$ the surface brightness is Σ_e. The word bulge, then, is the component of the surface brightness that is not in an exponential disk and which towards the center of the galaxy is comparable in brightness to the disk. Many bulges are better described by an exponential surface brightness distribution, but with a scale length significantly shorter than that of the disk.

It is worth pointing out that these functions hardly ever fit the full surface brightness profile of a galaxy in detail: disks in particular show departures from exponential from spiral arms, dust lanes, etc. These formulae have, however, been very successful in describing the large-scale structure of other spiral galaxies, and consequently there is an extensive literature on decomposition methods using these functional forms (see, for example, Schombert and Bothun 1987).

An alternative decomposition scheme, which does not rely on assumed functional forms for the disk and bulge, has been described by Kent (1986). In his method, the bulge and disk are decomposed into two profiles that have constant ellipticity[5] over radius. To illustrate Kent's method, I have plotted in Figure I the major-axis surface brightness profile of NGC 2841 at 2.2 μm (Terndrup et al. 1992). The profile was decomposed using an ellipticity of $e = 0.12$ for the bulge and $e = 0.45$ for the disk. Though this decomposition made no assumptions of the functional form of the profiles, the disk and bulge turn out to be close, respectively, to an exponential disk and an de Vaucouleurs bulge.

Regardless of the method of decomposition, it is important to remember that in other galaxies the terms bulge and disk are structural descriptions, and

---

[4]Additional components are typically needed to account for the high rotation speeds of the disks of spiral galaxies at large radii. These are imagined to be composed of "dark" (i.e., non-luminous) matter, and are not associated with any stellar population. Their function is to contribute gravity on large scales.

[5]The ellipticity of a light distribution is defined as $e = 1 - c/a$, where c/a is the axis ratio of the best-fitting ellipse at a given radius.

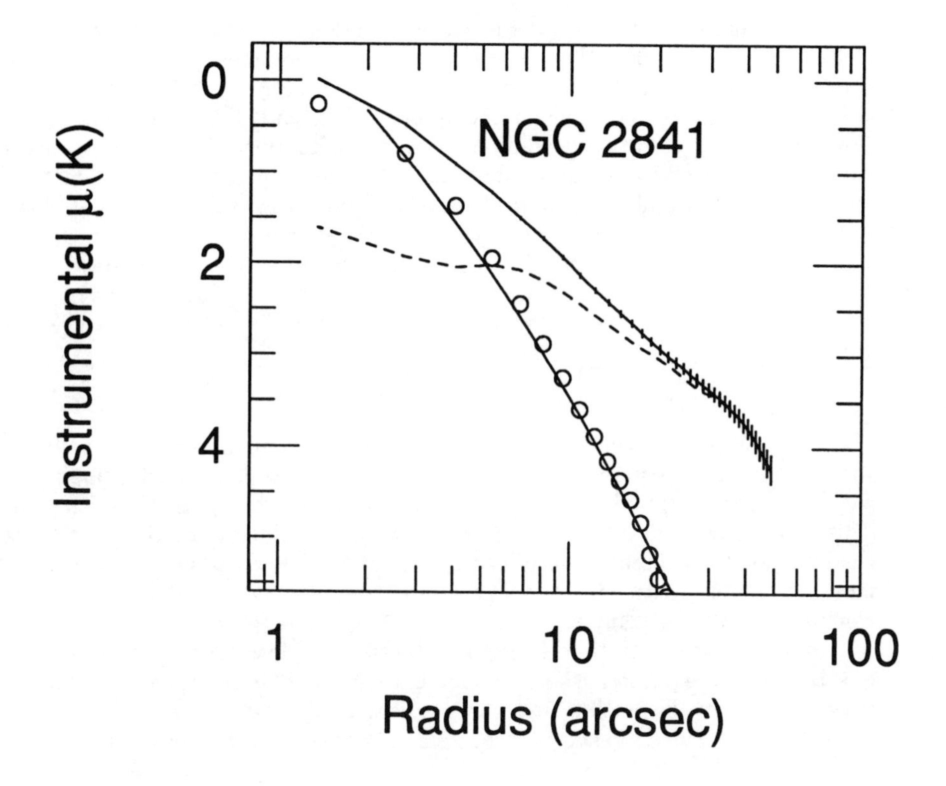

FIGURE I Major-axis surface brightness profile of NGC 2841 in the K band (2.2 μm). The observed profile is shown as a solid line with vertical bars, the length of which indicates photometric error; the magnitudes are with respect to the innermost measured point. The profile is decomposed into a disk (dashed line) and a bulge (open circles). The solid line through the bulge points is a de Vaucouleurs law fit to the bulge profile.

do not completely correspond to the way we use them in the study of the Galaxy's stellar populations. In other galaxies, the bulge component is significant only near the center of the light profile, and falls off rapidly at larger radii, though either a de Vaucouleurs profile or an exponential extends to infinity. The light of the bulge component at larger radii where the disk dominates may not necessarily correspond perfectly to the halo/thick disk that we see in the solar neighborhood.

Kent (1992) and collaborators have obtained the large-scale light distribution of our Galaxy from Spacelab data, and have provided a decomposition which in spirit is very much like applied to other spirals. Kent finds that the bulge can be fit by an exponential of scale length along the major axis of about 500 pc. This is about 10% the scale length of the disk (Lewis and Freeman 1989).

Stellar motions

Many stellar populations in the Milky Way are defined kinematically. Here I want to present a basic result: that the spatial distribution of a stellar population within the gravitational potential of a galaxy depends on its velocity dispersion. This discussion is drawn from the analysis of the kinematics of the globular cluster system by Frenk and White (1980) and follows the notation there. A more complete treatment of stellar kinematics may be found in Ivan King's contribution to this volume.

Stellar motions in the Galaxy's potential are described by the collisionless Boltzmann equation; the first moment of that equation gives the relation between velocity dispersion and the gradient of the mass distribution. For a spherically symmetric galaxy[6] we have in spherical polar coordinates

$$\frac{d}{dr}(\rho\sigma_r^2) + \frac{2\rho}{r}(\sigma_r^2 - \sigma_t^2) - \frac{\rho}{r}v_{\rm rot}^2 - \rho\frac{d\Phi}{dr} = 0, \qquad (3)$$

where $\Phi(r)$ is the gravitational potential, $v_{\rm rot}$ is rotation speed, ρ is the density distribution, and σ_r^2 and σ_t^2 are respectively the the squares of the radial and tangential velocity dispersions. In general, the velocity dispersion is a tensor, and can have unequal projections along the coordinate axes; for the spherical case being considered there are two dispersions, σ_r and σ_t. In the solar neighborhood we would use cylindrical coordinates and thus have σ_r (radial), σ_t (tangential), and σ_z (vertical).

The circular orbital velocity v_c and the Galaxy's mass distribution $M(r)$ are related through

$$v_{\rm c} = (GM/r)^{\frac{1}{2}} = \left(-r\frac{d\Phi}{dr}\right)^{\frac{1}{2}}. \qquad (4)$$

If we model the mass distribution as $\rho(R) \propto R^{-\nu}$, then $\nu = -d\ln\rho/d\ln r$, and equation (4) becomes

$$v_{\rm c}^2 = [\nu + 2\beta - 2]\sigma_r^2 + 2\sigma_t^2 + v_{\rm rot}^2, \qquad (5)$$

where β is $-d\ln\sigma_r/dr$.

---

[6]This is hardly true in the Galaxy, but the real situation will be qualitatively similar.

In equation (5), the terms ν, β, and v_c are constants at a given radius, depending on the mass distribution, so the variables are σ_r^2, σ_t^2 and v_{rot}. These all must add up so that the left hand side of the equation is a constant, so the orbital energy must be divided between rotation and random motions. A thermally cold population $\sigma^2 \approx 0$ must have a high rotation speed, and a hot population $\sigma \gg 0$ must be non-rotating.

POPULATIONS IN THE GALAXY

The local halo and thin disk

From the earliest large-scale surveys of stellar radial velocities (see Sandage 1986 for a review), it was evident that stars in the solar neighborhood fall into two groups. Stars in the first group have low velocities relative to the Sun in the direction of galactic rotation and a low velocity dispersion in the vertical direction ($\sigma_z \sim 20 - 30$ km sec^{-1}). The other group has large negative velocities and a vertical velocity dispersion of $\sigma_z \sim 100$ km s^{-1}. From equation (5), we can see immediately that the low velocity dispersion group must have a high rotation speed. Since $\sigma_z \ll v_{rot}$, this group was in a flattened system which could be identified with the plane of the Milky Way or the disks of other galaxies. That the average radial velocity of this group was small meant that the Sun was in this *disk* population. The second group had to have $v_c \sim 0$; with its high velocity dispersion the stars can achieve large distances from the galactic plane and form a *halo* population.

When it became possible to measure the chemical abundances of disk and halo stars, it was clear that the halo was metal poor, having values of the metallicity parameter[7] [Fe/H] < -1, while the disk had [Fe/H] > -1. The best modern data shows that the disk and halo are almost completely discrete from one another in metallicity and kinematics; this is shown in Figure II where I have plotted the correlation between metallicity and the V radial velocity (the component in the direction of the solar orbit around the Galaxy) from the sample of Carney et al. (1990a).

The scale heights perpendicular to the plane of various disk objects have been extensively studied (see, for example, the discussion in Mihalas and Binney 1981). Young disk objects, such as O and B stars, and their immediate progenitor gas and dust, have scale heights $z_h \leq 50$ pc. Stars that have older average ages, such as G-K-M dwarfs, have $z_h \sim 300 - 500$ pc. Though this could mean that the disk has gotten thinner over the lifetime of the Galaxy, the favored interpretation is that stars are scattered dynamically to larger vertical heights through gravitational interactions with molecular clouds.

In the halo, both RR Lyrae stars (Saha 1985) and globular clusters (Harris

---

[7]The metallicity parameter [Fe/H] is defined as the logarithmic abundance of iron compared to hydrogen, normalized to the solar value:

$$[\mathrm{Fe/H}] = \log\left[\frac{N(\mathrm{Fe})}{N(\mathrm{H})}\right] - \log\left[\frac{N(\mathrm{Fe})}{N(\mathrm{H})}\right]_{\odot} .$$

Thus a star with [Fe/H] $= -2$ would have 1% of the solar abundance of iron.

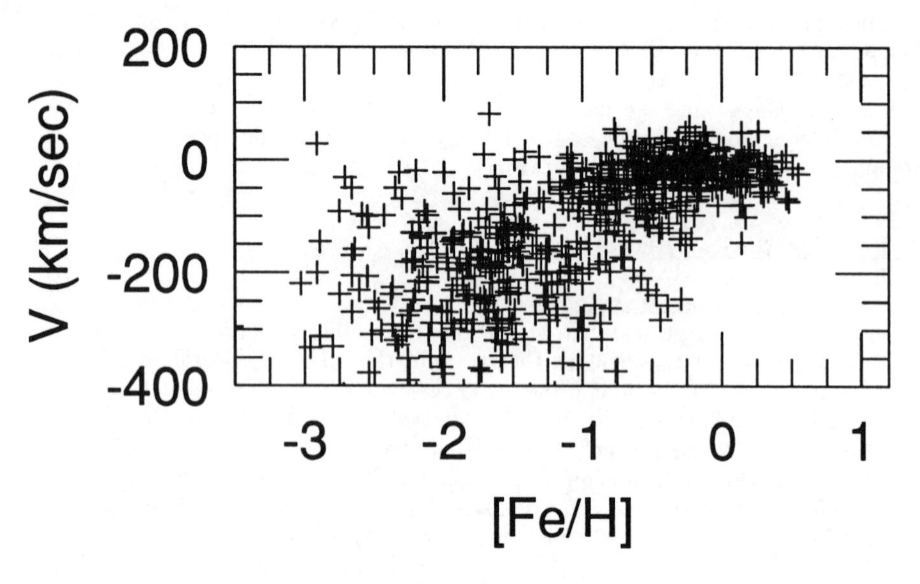

FIGURE II Relation between the V component of stellar velocities and metallicities for stars in the solar neighborhood. The velocity V is the component of orbital velocity in the direction of galactic rotation. Data are from Carney et al. 1990a).

1976) are distributed like $\rho \propto R^{-\nu}$, with $\nu = 3$ to 3.5. The high-velocity stars of the halo that are near the Sun have an anisotropic velocity dispersion, with $\sigma_r \approx 130$ km sec^{-1}, $\sigma_t \approx 100$ km sec^{-1}, and $\sigma_z \approx 90$ km sec^{-1} (Carney and Latham 1986, Norris 1986, Morrison et al. 1990). For a round gravitational potential, the halo would have an axial ratio of $c/a = 0.7$, but because of the disk the potential is flatter, and the halo would have $c/a \sim 0.3$ (Binney and May 1986). The halo is probably rounder than this outside the solar circle (Hartwick 1987).

Halo stars turn out to be uniformly old, since they lack main-sequence stars of high or intermediate mass. It was long thought that the dispersion of ages in halo stars was negligible, for two reasons. First, it was assumed that halo stars and globular cluster stars were members of the same population; though there is now strong evidence to the contrary, most measurements of the ages of the globular clusters indicated that the clusters were coeval to 1 Gyr or so. Secondly, if the Galaxy's halo was formed by a monolithic collapse of a large gas cloud, the only "natural" time scale was the gravitational free-fall time $t_{ff} = 1/(G\rho)^{1/2}$ (Eggen et al. 1962), which is only about 10^8 yr. Shuster and Nissen (1989), however, have called this assumption in to question. They obtained a sample of high velocity F stars near the Sun, and determined the metallicities and luminosities in their sample from $uvby\beta$ photometry. They derive an age spread in the halo of a few Gyr, about twice the typical error of measurement. This is in agreement with the latest data on age spreads in the globular cluster system, but possibly in conflict with the analysis of the chemical abundance patters in halo stars, which I will review later on.

The local thick disk

In the solar neighborhood, star counts and kinematically selected samples indicate the presence of another stellar component to the Galaxy besides the disk and halo, which is called the *thick disk.* It was characterized by Gilmore and Reid (1983) who showed that star counts in the Galaxy indicated the presence of a component having a scale height in the vertical direction of about 1 kpc, in contrast to (thin) disk tracers, which have scale heights of 350 pc or less. In the plane, the fraction of stars that are members of the thick disk is on the order of 2 – 5%; because of this, they contribute a significant fraction of the volume density of stars at distances of 1.5 – 2 kpc above the plane. Consequently, most studies of thick disk stars have investigated regions of space at that vertical height.

The metallicities of thick disk stars are on average about [Fe/H] = −0.7 (e.g., Friel 1987, 1988), intermediate between the metallicities of the disk and halo. Their kinematics, too, lie between halo and disk: the vertical velocity dispersion is ~ 45 km $\sec^{-1}$, and they have a slower rotation rate than the disk (seen as a mean lag of 30 – 50 km $\sec^{-1}$ behind the solar rotation). These characteristics are similar to the "Intermediate Population II" stars of the 1957 Vatican Conference.

Globular clusters

The globular clusters in the Galaxy have long been used as tracers of galactic structure, most notably when Shapley used their distribution and (then) crude distances to demonstrate that the Sun is not located near the center of the Galaxy. They are extremely useful as laboratories of stellar evolution, since with few exceptions the stars within a given cluster have nearly identical metallicities and a negligible dispersion in age. I have already mentioned the key role that globular clusters played in Baade's original definition of stellar populations. In this section I want to review the current status of the globular cluster system, drawing mainly from Armandroff's (1992) recent review.

Though there were persistent suggestions that the globular cluster system was not chemically or kinematically homogeneous, it was the analysis by Zinn (1985) that first demonstrated clearly that the properties of the globular clusters undergo an abrupt change at [Fe/H] = −0.8. The full distribution of globular cluster metallicities is bimodal, with peaks at [Fe/H] ≈ -1.6 and -0.5 and a minimum at [Fe/H] ≈ -0.9; the gaussian widths of these two groups are $\sigma = 0.33$ and $\sigma = 0.22$, respectively (Armandroff and Zinn 1988). According to Zinn (1985), clusters with [Fe/H] < -0.8 are in a non-rotating, only slightly flattened system with a high velocity dispersion; the radial density distribution is close to $\rho \propto R^{-3}$ (Harris 1976). Since this is similar to the field halo density distribution derived from the RR Lyraes and the metallicities are like those of halo objects near the sun, the metal-poor group of globular clusters has been called the *halo globular cluster system.*

The metal rich clusters ([Fe/H] > -0.8) are found near the plane in a rotating system, which has been called the *disk globular cluster system.* The scale height of this system is about 1100 pc (Armandroff 1989), significantly higher than the scale heights of disk objects. This combination of spatial distribution and metallicities does, however, resembles the local thick disk, leading to the suggestion that the disk globular clusters and the thick disk are part of the same system.

For a long time it was thought that the globular clusters were coeval to a level of 1 – 2 Gyr, but it is only within recent years that a definite dispersion of age within the globular clusters has been discerned (see the review by Bolte 1992). The first piece of evidence for an age spread was an analysis of the globular clusters NGC 288 and NGC 362. These are extreme examples of a "second parameter" pair – clusters that have similar values of [Fe/H] but have different horizontal branch morphologies, suggesting that there is a second parameter besides age which determines the distribution of stars along the horizontal branch[8]. It has proven rather difficult to determine the absolute ages of globular clusters by fitting theoretical stellar evolutionary models to even very accurate photometry of main sequence stars: even with reasonably accurate estimates for the distance or the reddening to a cluster, uncertainties in these parameters and systematic errors in photometric calibration propagate to an error in a cluster age of ± 2 Gyr or so. There have been, however, some recent studies that carefully determine the age *difference* among globular clusters via careful differential photometry of cluster main sequences. In the case of NGC 288/362, two independent studies (Bolte 1989; Green and Norris 1990) have each shown that the former cluster was the older by 2 – 3 Gyr. This has lent considerable support to the idea that age was the second parameter.

Lee (1992) has recently analyzed the distribution of stars on the horizontal branches of globular clusters, in an attempt to see whether there is any variation with location in the Galaxy of the age spread among the globulars. His method is to assume that age is the second parameter; his theoretical models then yield age differences at a given morphology from the distribution of stars along the horizontal branch Lee finds that the clusters within $R = 8$ kpc are all coeval, and on average about 2 Gyr older than those with $8 \leq R < 40$ kpc; clusters in this latter range of radius have an age spread of 3 – 4 Gyr.

Though the metal-poor clusters are identified with the halo, there have been few studies to see whether the two populations are similar in detail. The total mass of the globular cluster system is at most a few percent of the halo mass, so it is worth asking whether the halo clusters are a biased tracer of the halo. Recently, Suntzeff et al. (1991) have addressed this question by compiling metallicity estimates for RR Lyrae variables in both the field halo and in globular clusters. They find that the mean metallicity and dispersion for both the field variables and clusters are the same outside the solar circle, but inside the solar circle the field RR Lyraes are more metal rich than the globular cluster population. However, when the cluster variables are compared directly with the field variables, it is found that the metallicity distributions are the same. One interpretation is there is a gradient in age in the halo, with the field population of the halo a few billion years older than the outer halo. This is consistent with

[8] "First parameter" clusters exhibit the horizontal branch morphology that one would expect if metallicity alone were the determining factor: horizontal branch stars in metal-poor clusters lie mostly to the blue of the RR Lyrae instability strip, and to the red for metal-rich clusters.

Lee's (1992) analysis of the globular clusters.

The bulge

An operational definition of the *bulge population* is simple: it is the population we see in Baade's Window after allowance is made for the presence of foreground disk stars. In the following discussion I will be very brief, and will concentrate only on the characteristics that can be measured optically or in the infrared from the ground in the various low-absorption "windows" through which the bulge can be seen. In particular, I will ignore the voluminous data on the bulge IRAS sources.

Until the mid-1970's, detailed studies of the bulge were very difficult. With the advent of digital detectors on large telescopes, however, the bulge is now within reach for routine study. Whitford (1978) began the current active phase of research on the bulge when he demonstrated that the spectrum of the integrated light in Baade's Window is strong lined, quantitatively similar to the light of ellipticals and the bulges of other galaxies. Subsequent investigations up to the late 1980's (see the review by Frogel 1988) have determined the metallicity and age of bulge stars; at present there are many ongoing programs to characterize the bulge's kinematics and to constrain models of the formation of the bulge.

The metallicities of bulge K giants have been extensively studied since stars of any metallicity pass through this domain in old populations. Spectroscopic surveys by Whitford and Rich (1983) and Rich (1988) have shown that the mean metallicity of the bulge is very high ([Fe/H] $\sim +0.3$), and that there is a wide range of abundances present, from [Fe/H] $= -1.0$ to $+0.7$ or even higher. (The precise determination of the metallicity scale for such metal rich stars is very uncertain, but independently of the calibration about one third of the stars in Rich's sample are more strong-lined at a fixed color than anything in the solar neighborhood.) Recently, Geisler and Friel (1992) have confirmed this metallicity distribution using Washington photometry of the K giants. Rich's (1988) metallicity distribution is plotted as a generalized histogram[9] in Figure III.

Also shown on Figure III is the metallicity distribution of the RR Lyrae stars in Baade's Window, from the study of Walker and Terndrup (1991). The distribution is quite unlike that of the K giants; evidently the RR Lyrae variables are produced from only the extreme metal-poor tail of the K giant distribution. I will return to this point shortly when I discuss the age of the bulge, but given that metal-rich globular clusters generally populate their horizontal branches to the red side of the RR Lyrae instability strip, it is perhaps not surprising that only a subset of the bulge K giants are the progenitors of the RR Lyrae variables.

The bulge M giants have also been the subject of many studies, since they are relatively bright in the red and infrared, and are easily detected in grism surveys (e.g., Blanco et al. 1984, Blanco and Terndrup 1989, and references in these papers). The metallicity scale for M giants is even less certain than for K giants, since the atmospheres of such cool stars contain strong absorption from molecules and it is therefore hard to generate theoretical models for compari-

---

[9] A generalized histogram is constructed by summing unit-area Gaussians, centered on the x-axis value for a given contributor to the distribution; the width σ is set to equal the error in the x-value for that point.

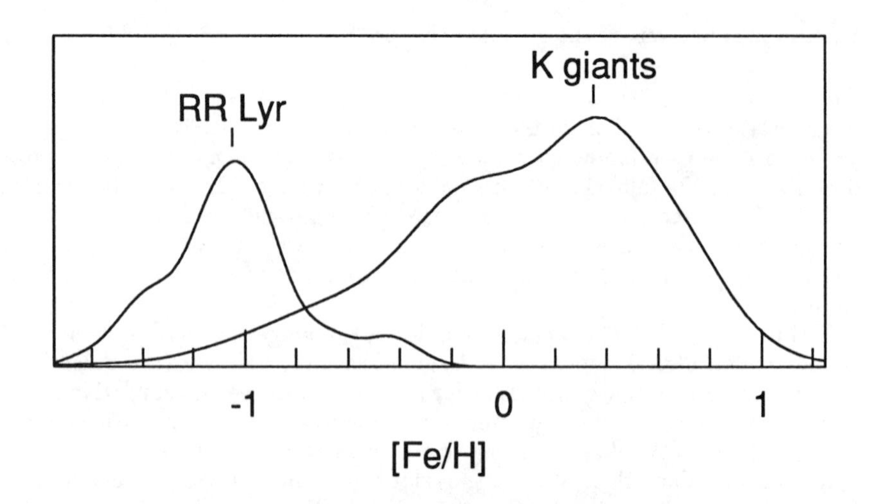

FIGURE III The abundance distribution for K giants in Baade's Window (Rich 1988) and of the RR Lyrae variables in the same field (Walker and Terndrup 1991). These are generalized histograms, and the vertical scale and relative numbers of stars under the two curves are arbitrary.

son to observed spectra. Bulge M giants have significantly stronger atomic and molecular absorption than do stars of the same temperature in the solar neighborhood or in globular clusters (Frogel and Whitford 1987; Frogel et al. 1990; Sharples et al. 1990; Terndrup et al. 1990, 1991); semi-quantitative analysis of the photometric and spectroscopic abundances of the Baade's Window M giants in all these papers yield a mean metallicity of [Fe/H] $\sim +0.2$ to $+0.4$, in agreement with mean metallicity of the K giants.

The bulge possesses a strong metallicity gradient, with the mean metallicity declining by several tenths of a dex from $R = 500$ pc to $R = 1500$ pc (Terndrup 1988; Frogel et al. 1990; Terndrup et al. 1990; Tyson 1991). To illustrate this, I reproduce as Figure IV the correlation of the strength of TiO absorption in bulge M giants as a function of infrared $J - K$ color for three groupings of stars along the minor axis of the bulge ($\ell = 0°$); this plot is from Terndrup et al. (1990). In the inner bulge fields ($b = -3°$ and $-4°$), the bulge M giants (plusses) have much stronger TiO absorption than do local stars (filled triangles). At higher galactic latitudes ($b = -10°$ and $-12°$), the bulge stars have weaker molecular absorption, much like that of the local stars. In this Figure, the transition between the inner and outer bulge is apparently rather abrupt. Tyson (1991) finds a similar result from his extensive study of K giant metallicities via Washington photometry in the same fields; he derives a mean abundance of [Fe/H] $= +0.3$ within $b \leq -10°$, falling by at least 0.75 dex at larger radii.

Unlike the K giants, the bulge M giants may not be an unbiased tracer of the bulge population, representing instead the metal-rich end of the metallicity distribution. There are two pieces of evidence which support this conclusion. The first, as shown in Figure V, is that the gradient in the area density of

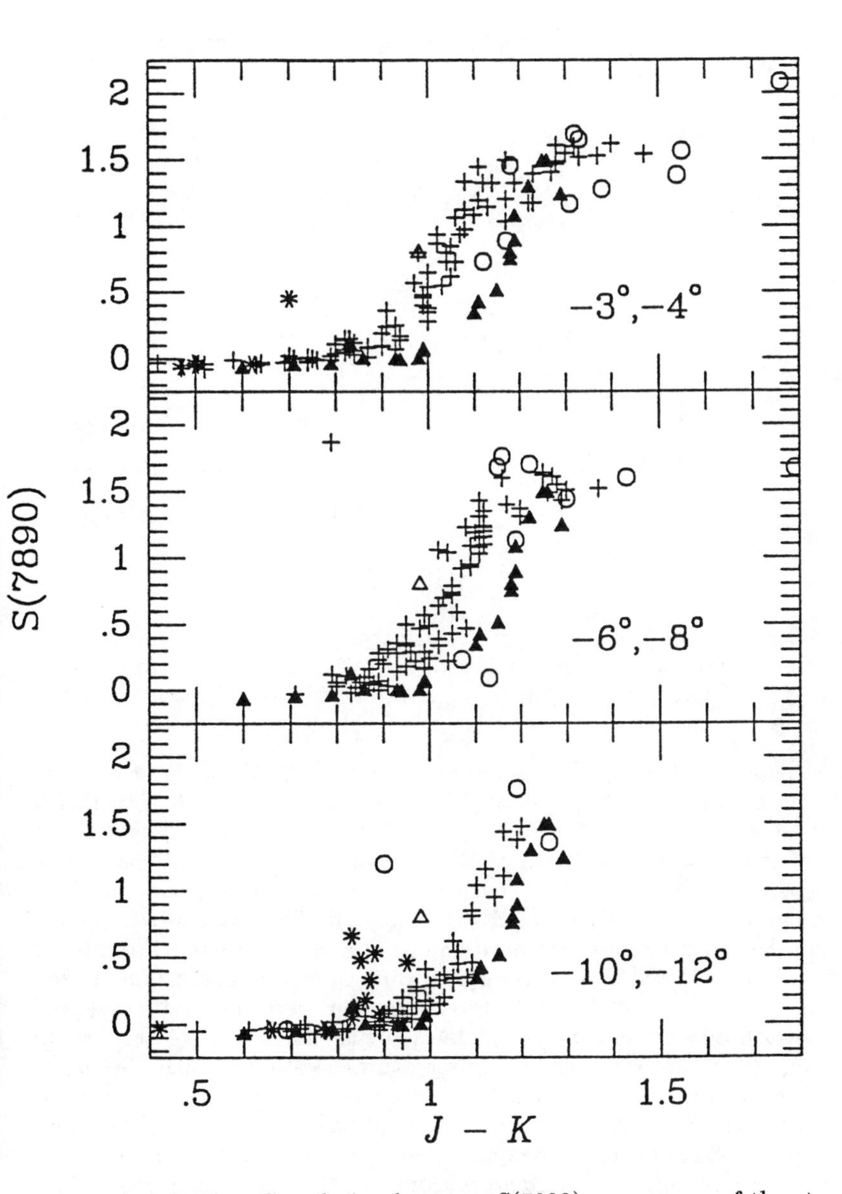

FIGURE IV Correlation between $S(7890)$, a measure of the strength of TiO absorption near 7890Å, and deredenned $J-K$ color. The data are for bulge fields along $\ell \approx 0°$, grouped into three pairs of fields: for $b = -3°$ and $-4°$ (top panel), $b = -6°$ and $-8°$ (center), and $b = -10°$ and $-12°$ (lower panel). Symbols are: *filled triangles,* solar-neighborhood M stars; *open triangle,* the nearby M dwarf Wolf 359; *plusses,* bulge M giants that are not long-period variables; *open circles,* bulge long-period variables; and *asterisks,* foreground M dwarfs seen in front of the bulge. Data are from Terndrup et al. 1990.

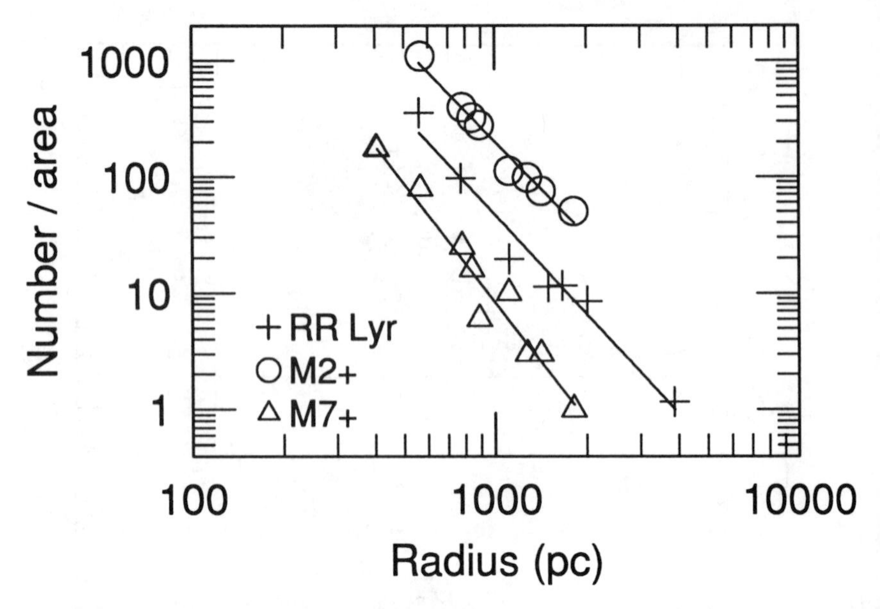

FIGURE V Surface density of various tracers of the bulge population. The plusses are for RR Lyrae variables, the open circles are for M giants of type M2 or later, and the open triangles are for giants of type M7 or later.

coolest M giants is significantly steeper than that of the K giants or early M giants (Blanco 1988; Terndrup 1988). The bulge K giants fall off with radius according to $\rho(R) \propto R^{-\nu}$, with $\nu = -3.5$, like that of the integrated light (Kent 1992) or of the halo, but the falloff of M giants of spectral type M7 or later is much steeper: more like $\nu = -4.3$. This is consistent with the observation of the metallicity gradient in the bulge: as the metallicity declines with increasing distance from the galactic center, the position of the giant branch shifts to warmer temperatures, so that relatively fewer giants are seen at very cool temperatures. This effect is also seen in globular clusters, where it is observed that metal-poor clusters never produce M giants, but metal-rich clusters such as 47 Tuc ([Fe/H] ≈ -0.7) do, though not with the cool temperatures seen in the bulge. Secondly, if the differences between bulge and local M giants in Figure IV are truly indicative of metallicity, and the M giants had the same metallicity distribution as the K giants which includes stars near [Fe/H] $= -1$, then there should be a considerable number of bulge M giants with weaker TiO absorption than in the solar neighborhood; this is not the case. From Figure IV it would seem that only K giants with [Fe/H] above solar can produce M giants. The exact distribution of metallicities of the M giants remains to be determined.

The age of the bulge can be determined directly from the main-sequence turnoff, which can be observed from the ground only for $|b| \geq 8°$ because of extreme image crowding. From optical color-magnitude diagrams, Terndrup (1988) derived a mean age of 11 – 14 Gyr, with little sign of any significant population of stars younger than 8 Gyr. Terndrup's mean age was derived

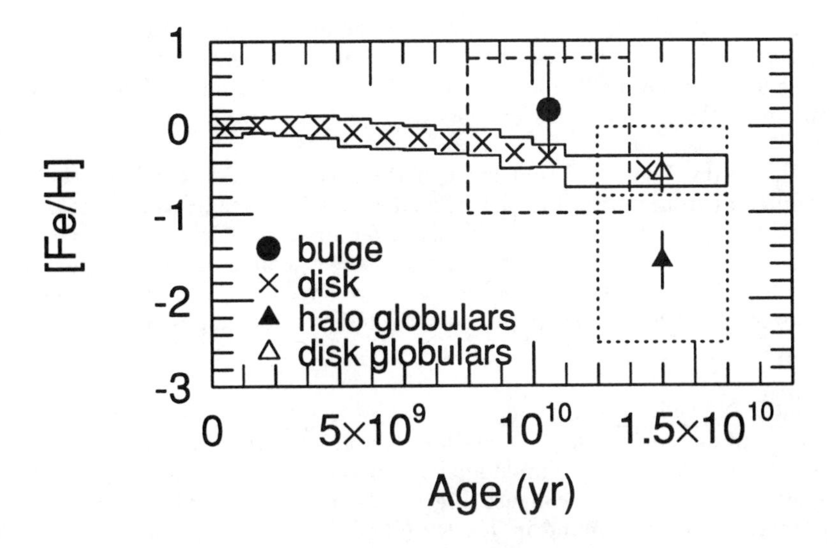

FIGURE VI Age metallicity relationship for the disk, globular clusters, and the bulge. See text for explanation.

assuming a distance to the bulge of $R = 7$ kpc; for the more traditional $R = 8$ kpc, the mean age is 9 – 12 Gyr. Very recently, the first observations of the main sequence turnoff in Baade's Window have been obtained with the WF/PC camera on the Hubble Space Telescope. Baum (1992) reports that the first analysis of the turnoff indicates a mean age of about 9 Gyr (for $R = 8$ kpc) more or less consistent with Terndrup's value; in any case the mean age is definitely several Gyr younger than the halo population. Furthermore, Baum states that the WF/PC turnoff suggests that the spread of ages in the bulge is significant.

COMPARING POPULATIONS: SIGNS OF HISTORY

In the remainder of this chapter, I will compare the stellar populations of the Galaxy, and draw a few conclusions about the the Galaxy's history. In particular, I would like to show that there is increasing evidence that the bulge and halo may be historically related to one another, with the halo proceeding the bulge. This "outside-in" picture contradicts Lee's (1992) "inside-out" model of the ages of RR Lyrae stars; that it does so indicates that there are still many details of stellar evolution that remain to be understood.

Age/metallicity relation

Figure VI summarizes available data on the ages and metallicities of stars in the halo, disk, and bulge. The age-metallicity relation for the disk is from Twarog (1980) and is shown as $\times$'s in boxes; the size of these boxes indicates the range of metallicities found at each age. Similar boxes, drawn as dotted lines, are shown for the bulge and for the halo and disk globular clusters; the extent of the box in the time direction qualitatively indicates either the uncertainty in the mean age or the possible age spread in these populations. The 1σ dispersion in abundance is shown as a vertical bar. The age of the bulge is from Terndrup (1988), and the mean age of the globular cluster systems is semi-arbitrarily set to 14×10^9 yr.

It is clear from Figure VI that the evolutionary history of the bulge/halo was very different from that of the disk. The disk has had steady star formation for the last 10^{10} yr, but has achieved only a factor of ~ 4 increase in metal abundance in that time. The halo and bulge, however, had an increase in metal abundance by a factor of several hundred in a few Gyr. On this plot, the bulge can be viewed as an extension of the halo to higher metallicities at (slightly) younger ages.

Simple models of chemical enrichment

I will now show that we can learn something about the evolutionary history of the Galaxy's populations by a simple analysis of their metallicity distributions. Throughout this discussion I will use the notation of Tinsley (1975); a more detailed treatment may be found in Tinsley (1980).

Imagine that a region of the Galaxy in which stars form and die has a total mass m divided between gas of mass m_g and stars of mass $m - m_g$. Define the gas fraction $\mu = m_g/m$ and the metal content of the gas as Z. There may be gas flowing into the system at a rate f (mass per unit time) with metallicity Z_f. Suppose stars are formed out of the gas at a rate ψ, and that these stars have a metal abundance of Z_*. Let R be the fraction of mass in each generation that is returned to the gas, and $y(1-R)$ be the fraction returned as newly synthesized metals. The term y is called the *yield.*

Depending on the relative values of f and y, the metal abundance of the gas can decrease, remain constant, or increase with time. Here we will consider only models that have the gas metallicity increasing monotonically with time, so that Z_* will increase steadily also. Thus Z_* becomes an axis of *time*; at each moment stars of metallicity Z_* are making the metals for the next generation of stars, which will have higher values of metallicity.

Let S be the total mass of stars born up to time t; contained in S are both the mass in the stars at any moment and the mass returned to the interstellar medium as stars die. For models with monotonically increasing abundances, the quantity S can also be expressed as the fraction of stars with metallicity $\leq Z_*$. For simplicity, assume that star formation produces only two types of stars – those that live forever and those that return metals to the gas through nucleosynthesis immediately after they are born. This last assumption is termed *instantaneous recycling.*

With these definitions, the basic equations for enrichment are:

$$\begin{aligned} dS/dt &= \psi, & (6) \\ dm/dt &= f, & (7) \\ dm_g/dt &= -(1-R)\psi + f, & (8) \\ d(Zm_g)/dt &= -Z_*(1-R)\psi + y(1-R)\psi + Z_f f. & (9) \end{aligned}$$

We can solve these equations to derive the ratio S/S_1 as a function of Z/Z_1, where the subscripts 1 denote the present values of these quantities. These can be parameterized either in terms of the yield y or the gas fraction μ. These are related by $y = Z_1/(\ln \mu^{-1})$.

There are two cases to consider. In the first, assume $f = 0$, $Z_* = Z$, y and R are constants, and initially the gas was unenriched: $Z(0) = 0$. Then it is straightforward to show

$$\frac{S}{S_1} = (1 - \mu^{\xi})/(1 - \mu_1), \tag{10}$$

where $\xi = Z/Z_1$. Because $f = 0$, this is called the *closed-box* model.

In the second case, assume that the gas initially begins with a floor metallicity $Z(0) = Z_0$. The resulting formula for S/S_1 is the same, except that $\xi = (Z - Z_0)/(Z_1 - Z_0)$. This is called an initial-enrichment model.

In Figure VII are plotted the cumulative metallicity distributions S/S_1 in the bulge (top panel) and in the disk (bottom). The horizontal axis is plotted as [Fe/H] instead of Z/Z_1. Shown as a dashed line are closed-box models for the metallicity distributions; for the disk I have also plotted an inital-enrichment model (dotted line).

The lower panel of Figure VII illustrates a classic problem in stellar populations, the "G-dwarf" problem. The metallicity distribution of the disk is derived from stars that have main-sequence lifetimes longer that the age of the Galaxy, i.e., G-K-M dwarfs. Closed box models with Z_1 equal to the maximum observed metallicity in the disk (about twice solar) and with reasonable values of the current gas fraction ($\mu \sim 0.05$ is plotted here) always predict many more metal poor stars than are observed in the disk. The models need to have a significant number of metal-poor stars to generate the metals needed for the next generation; the disk does not have the requisite number of metal-poor stars.

Alternatively, we can try initial-enrichment models that have the disk's metal abundance starting from some floor value. A specific case, in which $Z_0 = 0.1Z_\odot$ is shown as a dotted line in the lower panel of Figure VII. This is somewhat of an improvement, but not a very good fit. Various alternative schemes have been tried, as reviewed by Tinsley (1975, 1980), but as with the models presented here they fail to explain the distribution of metals in the disk when it is treated as an isolated system.

In the bulge, however, the closed-box model[10] works well (top panel of Figure VII). The specific model is one with low current gas fraction $\mu_1 = 0.005$ (i.e., the bulge gas was all used up to make stars); this corresponds to a yield of twice the solar abundance. That the simple model works suggests that the bulge could have been self-enriched: starting with a low initial metallicity, the gas in what is now the bulge went through generations of star formation in relative isolation until the gas was all consumed.

Now because the abundance in the halo of the Galaxy is very low (Figure II), star formation in the halo did not proceed very far before it stopped: if the halo gas had been all used up there would be high abundances in the halo like those of the bulge. The simplest scenario for the formation of the halo is that star

---

[10] Rich (1990) discusses this in more detail, summarizing several model fits to the bulge metallicity distribution.

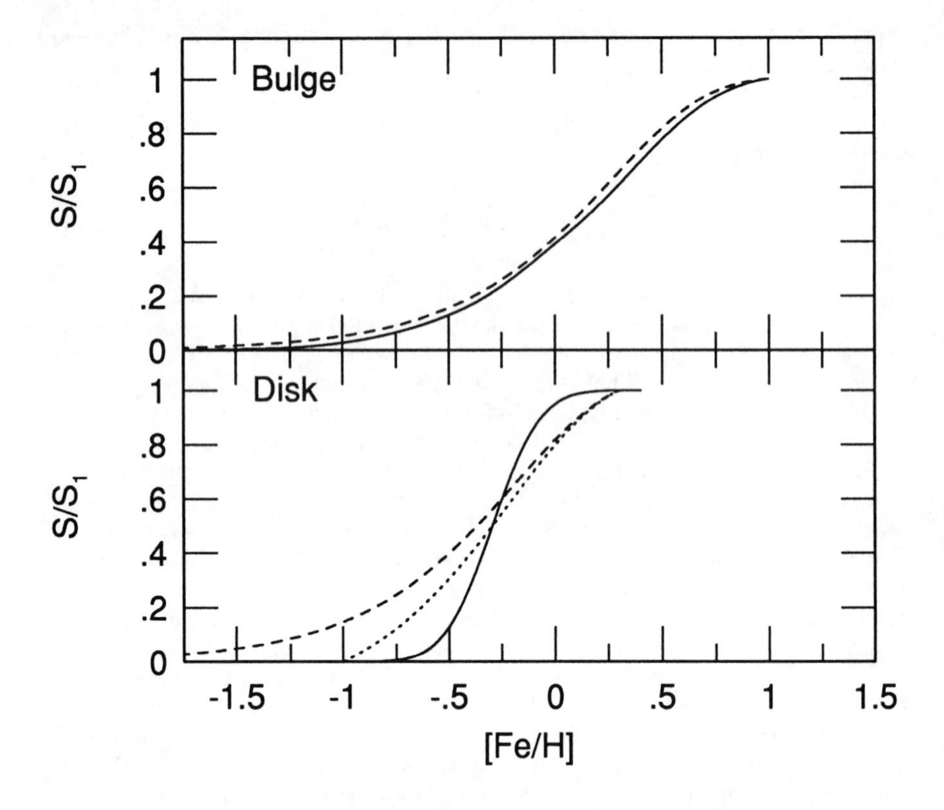

FIGURE VII Models for the abundance distribution in the disk and bulge. In both panels, the solid line shows the observed distribution: the bulge data are from Rich (1988), and the disk from Pagel and Patchett (1975). The dashed line is for a closed-box models in which Z_1 was taken to be the largest metallicity in the observed distributions. An initial-enrichment model is also shown for the disk (dotted line).

formation began in an initially unenriched gas for a short period, then the gas was lost from the halo. Where did the gas go? Carney et al. (1990b) argue that the most natural sink for the lost halo gas was the bulge. They calculate that the total mass of the stellar halo is $1.5 \times 10^9 M_\odot$, and from chemical enrichment arguments like the ones presented here conclude that only 1/30th of the original halo gas was converted into stars. Thus the expected mass of the bulge should be about equal to the mass of the lost halo gas, or about $4 \times 10^{10} M_\odot$; this is within a factor of a few of the observed mass (Blanco and Terndrup 1989). Thus the high metallicities of bulge stars does not *a priori* rule out an evolutionary connection between bulge and halo: it was only in the bulge that star formation went to completion and used up the majority of the gas, a process that necessarily produces high abundances.

Abundance patterns in the Galaxy

In the previous discussion we were able to draw a few interesting conclusions about the nature of chemical enrichment in the solar neighborhood despite making many simplifying assumptions. I would now like to discuss how one assumption in particular – that stellar metal abundance can be described by a single number Z – is not true in a way that tells us a considerable amount about the *time scale* for chemical enrichment. This subject is extensively reviewed by Tinsley 1980, Spite and Spite 1985, Gilmore et al. 1989, and Wheeler et al. 1989.)

Alhough not all the details of supernovae explosions are known, most models indicate that there are three basic modes of chemical enrichment:

- Very massive type II supernovae ($M > 30M_\odot$) produce ejecta with large amounts of O, with very little C, Fe, or r-process elements. (The r-process elements are those produced by rapid neutron capture; some r-process elements are Eu, Dy, Gd.) Stars of this mass have very short lifetimes, on the order of 10^7 yr, so that these stars will be the first ones to contribute their newly synthesized metals to the interstellar medium and ultimately to the next generations of stars.

- More typical type II supernovae, those with $30M_\odot > M > 10M_\odot$, are expected to produce ejecta rich in the r-process elements and the α-elements, as well as producing some Fe. (The α-elements are so named because they are produced from repeated capture of α particles; O, Ca, Si, Mg, and Ti are α-elements.) The main sequence lifetimes of these stars are from 10^8 to 10^9 yr.

- Intermediate-mass type I supernova ($10M_\odot > M > 1M_\odot$) produce more iron than type II supernovae. The lifetimes of these stars are in excess of 10^9 yr, so chemical enrichment by this process takes place over considerably longer time scales than do the other processes.

Consider the simplified case of the relative abundances of oxygen and iron. These have different *rates* of chemical enrichment: oxygen production from massive stars takes place over a time scale $t < 10^9$ yr, while iron production from lower mass stars is important only for $t > 10^9$ yr. (These numbers depend on the details of supernova theory, and are still uncertain to a factor of at least two, but that uncertainty is not important in this discussion.)

Now the halo and disk differ not only in mean metallicity, but in the ratio of oxygen to iron. Stars with [Fe/H] < -1, which are primarily in the halo, have [O/Fe] $\sim +0.5$, while disk and thick disk stars with [Fe/H] > -1 have [O/Fe] $\rightarrow 0$ as [Fe/H] $\rightarrow 0$ (Wyse and Gilmore 1988, Bessell 1991). From the above arguments, it would then seem that the halo stars received their metals over a time scale dominated by type II events, and thus was enriched in a short period of time ($< 10^9$ yr). The disk, on the other hand, was enriched over a sufficiently long time that both type I and type II supernova contributed.

Until recently, there had been no studies of the chemical abundance patters in the nuclear bulge, mostly because even the K giants are rather faint for the high-resolution spectroscopy needed to measure element ratios. At present, spectra of a very small sample (12 stars) have been obtained in the bulge by McWilliam and Rich (1992). They find that several of their stars exhibit a mild enrichment of the α-elements, on the order of $[\alpha/\mathrm{Fe}] = +0.2$ to $+0.3$. Perhaps most interesting is that they derive very high abundances of the r-process element Eu. The two results together suggest that the bulge's chemical enrichment may have been rapid, like that of the halo, and further strengthens the arguments from the simple chemical evolution models that the bulge and halo may be related to one another.

Bulge kinematics

I would like to summarize the current information on bulge kinematics, and draw a few conclusions about the structure of the bulge and how stars of different metallicities may differ in spatial distribution. Rich (1990) and Sharples et al. (1990) have presented similar discussions in their respective studies of the K and M giants in Baade's Window.

There is some very preliminary evidence from Rich (1990) and Spaenhauer et al. (1992) that the velocity dispersion of bulge objects is a function of metallicity. Their data are summarized in Table I. Rich (1990) presented velocities

TABLE I Line-of-sight velocity dispersions in Baade's Window.

Source	σ_r	σ_ℓ	σ_b	N
Rich (1990) K giants	105 ± 11	...	...	53
Rich, [Fe/H] < -0.3	126 ± 22	...	...	16
Rich, [Fe/H] $\geq +0.3$	92 ± 14	...	...	21
Sharples et al. (1990) M5+ giants	113 ± 11	...	...	225
Spaenhauer et al. (1992) all stars	...	115 ± 4	100 ± 4	429
Spaenhauer et al. with [Fe/H] $\geq +0.0$	101 ± 15	118 ± 14	58 ± 9	34

of 53 K giants in Baade's Window, and derived a line-of-sight radial velocity dispersion $\sigma_r = 105 \pm 11$ km sec^{-1} for his full sample. He then divided his sample into bins of metallicity, and showed (though with small number statistics) that the radial velocity dispersion for the most metal rich stars ([Fe/H $\geq +0.3$) was 92 ± 14 km sec^{-1}, much lower than 126 ± 22 km sec^{-1} for stars with [Fe/H]

< -0.3. This effect was also seen in the proper motion data of Spaenhauer et al. (1992), who measured relative proper motions in the ℓ and b directions for over 400 stars in Baade's Window. (The proper motions are relative to the average of the sample because there are no background galaxies or proper motion standard visible in Baade's Window; their study therefore is a measure of the *dispersion* in the line-of-sight proper motion.) When they add radial velocities from Rich, Spaenhauer et al. find that the velocity dispersions in Baade's Window are nearly isotropic in projection, meaning the line-of-sight velocities are nearly equal: $\sigma_r \approx \sigma_\ell \approx \sigma_b$. When they use the small number of Rich's most metal-rich stars that are also in their sample, Spaenhauer et al. show that the vertical velocity dispersion $\sigma_b \sim 60$ km sec^{-1} is smaller than the tangential velocity dispersion σ_ℓ. If the lower velocity dispersion for the metal-rich stars is real, what would be the relative rotation speeds and spatial distribution of the metal-rich stars compared to the general bulge population?

Bulges of galaxies like our own (between Hubble type Sb and Sc) are *not* self-gravitating bodies; the gravitational potential in the bulge has a very significant contribution from the disk. Consequently, the proper way to explore kinematics in the bulge is to generate a mass model, as Kent (1992) has done from the Galaxy's K-band light distribution, and correctly solve the equations relating kinematics and the density distribution. On the other hand, we can estimate how stars of different kinematics will be arranged in the bulge by making use of eq. (5). The circular velocity v_c is a constant at a given radius, so for two populations in the same gravitational potential

$$1 = \frac{[\nu_1 + 2\beta_1 - 2]\sigma_{r,1}^2 + 2\sigma_{t,1}^2 + v_{\mathrm{rot},1}^2}{[\nu_2 + 2\beta_2 - 2]\sigma_{r,2}^2 + 2\sigma_{t,2}^2 + v_{\mathrm{rot},2}^2} \tag{11}$$

From Figure V, we saw that $\nu \sim 3.5$ for bulge K giants and RR Lyraes and $\nu \sim 4.2$ for late M giants. In Figure VIII are plotted measures of the line-of-sight velocity dispersion from several published papers and from my own work, with is mostly in preparation. These data are for K and M giants both along the minor axis of the bulge and for some fields several degrees away. The correlation between σ and galactocentric distance R is quite tight, and gives $\beta = 0.4$. There does not seem to be any variation in the value of β between the K and M giants.

Let the subscript 1 refer to the general bulge population, for which $\sigma_{r,1} =$ 105 km sec^{-1} (Rich 1990) and $\sigma_t \approx \sigma_r$. Direct measures of the rotation speed for bulge objects give $v_c \approx 80$ km sec^{-1} (Kinman et al. 1988, Menzies 1990, Minniti et al. 1992); the bulge would also have to have this value of v_c to have the observed axial ratio $c/a = 0.7$ (Kent 1992, Davies et al. 1983). This value of 80 km sec^{-1} is very interesting, since it is the expected rotation speed if the bulge was the inner part of the halo, which has $v_{\mathrm{rot}} \sim 10$ km sec^{-1} near the Sun (Carney et al. 1990b); the higher velocity is just from conservation of angular momentum, implying that the bulk of bulge stars formed in a dissipation-free collapse.

Assume the velocity dispersion for the metal-rich K giants is also isotropic; if these stars have $\nu = 4.2$ as for the late M giants, then from eq. (11), we have $v_{c,2} = 115$ km sec^{-1}. For this combination of σ and v_c, the metal-rich stars would have to be in a flatter distribution than the majority of bulge stars, having $c/a \sim 0.5$. Furthermore, the low vertical velocity dispersion of these stars

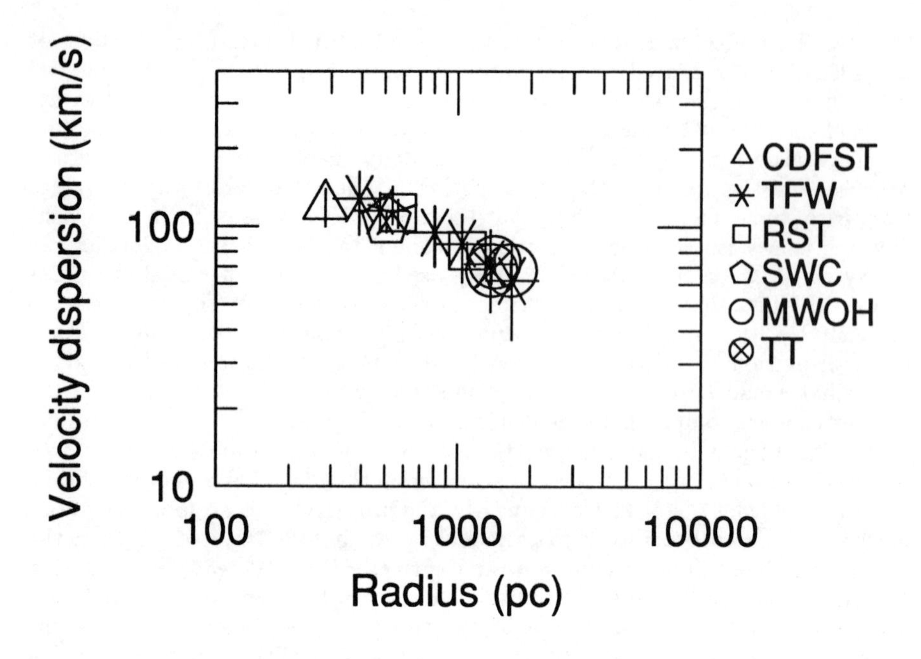

FIGURE VIII Line-of-sight velocity dispersion in the bulge. The sources for the points are: CDFST: Carr et al. 1992; TFW: Terndrup et al. 1992; RST, Rich et al. 1992; SWC: Sharples et al. 1990, MWOH: Minniti et al. 1992; TT: Tiede and Terndrup 1992.

may indicate that there was some dissipation during the formation of the most metal-rich stars in the bulge, as happened in the disk.

From these simple comparisons of populations, we have some reasonably interesting results: The bulge/halo system was probably formed rapidly ($\sim 1 - 2 \times 10^9$ yr), during which the bulge became more highly flattened. This very simple picture hides many details and is certainly only a first-order answer. A full understanding of the formation of the bulge probably requires a complicated dynamical model, including the possibility of satellite accretion, bar formation, gas flows, etc.

ACKNOWLEDGMENTS

I would like to thank Roberta Humphreys and the Department of Astronomy of the University of Minnesota for their generous hospitality during my visits. My thanks also to the graduate students, with whom I had many stimulating conversations about the Galaxy, politics, sports, and life.

REFERENCES

Armandroff, T. E. 1989, *AJ*, 375

Armandroff, T. E. in *The Globular Cluster - Galaxy Connection, Eleventh Santa Cruz Summer Workshop*, in press

Armandroff, T. E., and Zinn, R. 1988, *AJ*, **96**, 92

Baade, W. 1944, *ApJ*, **100**, 137

Baade, W. 1951, *Publ. Obs. Univ. Mich.*, **10**, 7

Baum, W. 1992, private communication

Bessell, M. S., Sutherland, R. S., and Ruan, K. 1991, *ApJ*, **383**, L71

Binney, J., and May, A. 1986, *MNRAS*, **218**, 743

Blaauw, A. 1965, in *Galactic Structure*, eds. A. Blaauw and M. Schmidt, (Chicago: University of Chicago Press), p. 435.

Blanco, V. M. 1988, *AJ*, **95**, 1400

Blanco, V. M., McCarthy, M. F., and Blanco, B. M. 1984, *AJ*, **89**, 636

Blanco, V. M., and Terndrup, D. M. 1989, *AJ*, **98**, 843

Bolte, M. 1989, *AJ*, **97**, 1688

Bolte, M. 1992, *PASP*, **104**, 794

Carney, B., and Latham, D. W. 1986, *AJ*, **97**, 423

Carney, B. W., Aguilar, L., Latham, D. W., and Baird, J. B. 1990a, *AJ*, **99**, 201

Carney, B. W., Latham, D. W., and Laird, J. B. 1990b, *AJ*, **99**, 572

Carr, J., DePoy, D., Frogel, J. A., Sellgren, K., and Terndrup, D. M. 1992, in preparation

Davies, R. L., Efstathiou, G., Fall, S. M., Illingworth, G., and Schechter, P. L. 1983, *ApJ*, **266**, 41

de Vaucouleurs, G. 1948, *Ann. d'Astrophys*, **11**, 247

Eggen, O. J., Lynden-Bell, D., and Sandage, A. 1962, *ApJ*, **136**, 748

Frenk, C. S., and White, S. D. M. 1980, *MNRAS*, **193**, 295

Friel, E. D. 1987, *AJ*, **95**, 1727

Friel, E. D. 1987, *AJ*, **93**, 1388

Frogel, J. A. 1988, *ARA&A*, **26**, 51

Frogel, J. A., and Whitford, A. E. 1987, *ApJ*, **320**, 199

Frogel, J. A., Terndrup, D. M., Blanco, V. M., and Whitford, A. E. 1990, *ApJ*, **353**, 494

Geisler, D., and Friel, E. D. 1992, *AJ*, **104**, 128

Gilmore, G., Wyse, R. F. G., and Kuijken, K. 1989, *ARA&A*, **27**, 555

Gilmore, G., and Reid, N. 1983, *MNRAS*, **202**, 1025

Green, E. M., and Norris, J. E. 1990, *ApJ*, **353**, L17

Harris, W. E. 1976, *AJ*, **81**, 1095

Hartwick, F. D. A. 1987, in *The Galaxy*, eds. G. Gilmore and B. Carswell, Dordrecht: Reidel, p. 281

Kent, S. M. 1986, *AJ*, **91**, 1301

Kent, S. M. 1992, *ApJ*, **387**, 181

King, I. R. 1971, *AJ*, **83**, 377

Kinman, T. D., Feast, M. W., and Lasker, B. M. 1988, *AJ*, **95**, 804

Lee, Y.-W. 1992, *PASP*, **104**, 798

Lewis, J. R., and Freeman, K. C. 1989, *AJ*, **97**, 139

McWilliam, A., and Rich, R. M. 1992, in preparation

Menzies, J. W. 1990, in *Bulges of Galaxies*, eds. B. Jarvis and D. Terndrup, *ESO Conf. Ser.,* **35**, 115

Mihalas, D., and Binney, J. 1981, *Galactic Astronomy: Structure and Kinematics*, (San Francisco: Freeman), p. 249*ff*

Minniti, D., White, S. D. M., Olszewski, E. W., and Hill, J. M., 1992, *ApJ*, **393**, L47

Morgan, W. W., and Osterbrock, D. E. 1969, *AJ*, **74**, 515

Morrison, H. L., Flynn, C., and Freeman, K. C. 1990, *AJ*, **100**, 1191

Mould, J. R. 1982, *ARA&A*, **20**, 91

Norris, J. 1986, *ApJS*, **61**, 667

Pagel, B. E. J., and Patchett, B. E. 1975, *MNRAS*, **172**, 13

Rich, R. M. 1988, *AJ*, **95**, 828

Rich, R. M. 1990, *ApJ*, **362**, 604

Rich, R. M., Sadler, E. M., and Terndrup, D. M. 1992, in preparation
Saha, A. 1985, *ApJ*, **289**, 310
Sandage, A. 1986, *ARA&A*, **24**, 421
Schombert, J. M., and Bothun, G. D. 1987, *AJ*, **93**, 60
Sharples, R., Walker, A., and Cropper, M. 1990, *MNRAS*, **246**, 54
Shuster, W., and Nissen P. 1989, *A&A*, **222**, 69
Spaenhauer, A., Jones, B. F., and Whitford, A. E. 1992, *AJ*, 103, 297
Spite, M., and Spite, F. 1985, *ARA&A*, **23**, 225
Suntzeff, N. B., Kinman, T. D., and Kraft, R. P. 1991, *ApJ*, **367**, 528
Terndrup, D. M. 1988, *AJ*, **96**, 884
Terndrup, D. M., Frogel, J. A., and Whitford, A. E. 1990, *ApJ*, **357**, 453
Terndrup, D. M., Frogel, J. A., and Whitford, A. E. 1991, *ApJ*, **378**, 742
Terndrup, D. M., Frogel, J. A., and Wells, L. A. 1992, in preparation
Tiede, G., and Terndrup, D. M. 1992, in preparation
Tinsley, B. M. 1975, *ApJ*, **197**, 159
Tinsley, B M. 1980, *Fund. Cosmic Phys.*, **5**, 287
Twarog, B. A. 1980, *ApJ*, **242**, 242
Tyson, N. D. 1991, thesis, Princeton University
Walker, A. R., and Terndrup, D. M. 1991, *ApJ*, **378**, 119
Wheeler, J., C., Sneden, C., and Truran, J. W. 1989, *ARA&A*, **27**, 279
Whitford, A. E. 1978, *ApJ*, **226**, 777
Whitford, A. E., and Rich, R. M. 1983, *ApJ*, **274**, 723
Wyse, R. F. G., and Gilmore, G., 1988, *AJ*, **95**, 1404
van den Bergh, S. 1975, *ARA&A*, **13**, 217
Zinn, R. 1980, *ApJS*, **293**, 424

The Minnesota Lectures on Clusters of Galaxies and Large-Scale Structure
ASP Conference Series, Vol. 39, 1993
Roberta M Humphreys (ed.)

THE GALACTIC HALO

KENNETH JANES
Boston University, Astronomy Department, 725 Commonwealth Ave., Boston, MA 02215

ABSTRACT The oldest observable objects in the galaxy are the stars of the galactic halo. Consequently, the properties of the halo stars, their compositions, kinematics, distributions and ages, should provide us with critical information about the earliest stages of the development of the galaxy. This review is a summary of those properties, based on a definition of the stellar halo.

1. INTRODUCTION

The halo of our galaxy is almost invisible, and if we look at edge-on spiral galaxies, we do not even notice that they have halos. So the first galactic astronomer, Sir William Herschel did not see anything like a halo, nor was there a halo in the so-called "Kapteyn Universe", a model for the galaxy dating from the beginning of this century.

The halo of the galaxy is generally taken to consist of the globular clusters along with those stars in the field that kinematically and chemically resemble those in the globulars. One can be more inclusive by adding the stars of the central bulge together with the globular clusters and related stars to make up a single entity known as the "galactic spheroid". Finally, in addition to the stars, and clusters of the stellar halo, there is the "dark matter halo", whose existence in some form now seems incontrovertible. One of the aims of these lectures is to develop a clearly-stated, self-consistent definition of the *stellar halo* and to describe the relationship between the stellar halo and other stellar and non-stellar components of our galaxy. A brief historical introduction will direct the way to a reasonable, consistent definition of the halo.

It was Harlow Shapley (1918a) in his work on the distribution of globular clusters who found the first real hint to a population of galactic stars external to the galactic disk. Shapley's work is known primarily because he recognized the existence of an identifiable group of stars that are *not* distributed in a disk-like way. He does not mention the word halo in his early papers, but he does say (Shapley, 1918b) that "The general system of clusters appears to be somewhat ellipsoidal..." This ellipsoidal distribution of the globulars was essential to his determination of the distance to the galactic center.

A second line of evidence for a stellar halo came from the study of stellar velocities. By the end of the 19th century, people had discovered some stars of unusually large velocity with respect to the Sun. The velocity distribution of these stars was highly asymmetric, directed primarily toward one galactic hemisphere. Oort (1922) showed that the velocity distribution of stars with total velocity less than 63 km/sec was symmetric about zero, but that the high velocity stars had a highly asymmetric distribution. This distinction led to the concept of the two "star streams" – one stream moving more or less at right angles to the direction of the galactic center and the other moving principally in the radial direction to the galactic center. Oort (1927) later confirmed Linblad's (1925) theory that the "high-velocity stars" are moving in primarily radial orbits about the galactic center, whereas the ordinary, low-velocity stars are revolving in nearly circular orbits about the center of the galaxy.

To the combination of kinematic and distributional differences in the two groups of stars was added the realization by Baade (1944) that the stars in the central bulge of M31 are all old, whereas many of those in the spiral arms are young. He made the first use of the concept of *stellar populations* – families of stars with definable properties that separate them from other stars. Thus (in Baade's view) the old, red stars in the M31 central bulge, the centrally concentrated globular clusters, and the radially-moving high velocity stars are drawn from the same population of stars.

Roman's (1952) classic paper on the compositions of the high-velocity stars added another dimension to the defining properties of stellar populations. She found that the high-velocity stars (i.e. halo stars) are metal-poor relative to the bulk of the stars in the solar neighborhood.

All of these threads were woven together at the 1957 Vatican "Conference on Stellar Populations" (O'Connell, 1958), in which the participants defined as many as 5 stellar populations, ranging from the Halo Population II to the extreme Population I. The defining properties of these populations were specified in terms of the types of stars found, their kinematics, distributions, ages, and chemical compositions.

The final step in the process of developing a concept of the galactic halo was the use of stellar kinematic data to derive a scenario for the formation of the galaxy. Eggen, Lynden-Bell and Sandage (1962, hereafter ELS) showed that the kinematics and abundances of stars could be explained by a model in which the halo stars (Population II) formed from a primordial cloud as it collapsed in a state of near free-fall. As the proto-galaxy continued to collapse, supernovae from the first generation of stars enriched the remaining gas with heavy elements. In the later stages of the collapse, angular momentum, and perhaps the increased gas density, slowed the collapse leading to the formation of the galactic disk. The key demonstration of the ELS hypothesis is the correlation they found between the orbital eccentricity of stars and their ultraviolet excesses, $\delta(U-B)$, which is correlated with metallicity (see figure 1).

This scenario for the formation of the galaxy, with its implications of a continuous, smooth transition in properties between the halo stars formed from freely falling gas and the disk stars from a more-or-less equilibrium gaseous disk, has dominated our view of the halo for 30 years. It is only in the past few years that an alternative hypothesis has begun to gain acceptance, the Searle and Zinn (1978) suggestion that the halo formed from mergers of smaller fragments.

The purpose of this brief historical introduction is to set up a basis for a definition of the halo that is consistent with the historical background as well as meaningful for comparisons with modern theory. We can see from the preceding discussion that the idea of the galactic halo is based on the separation of stars

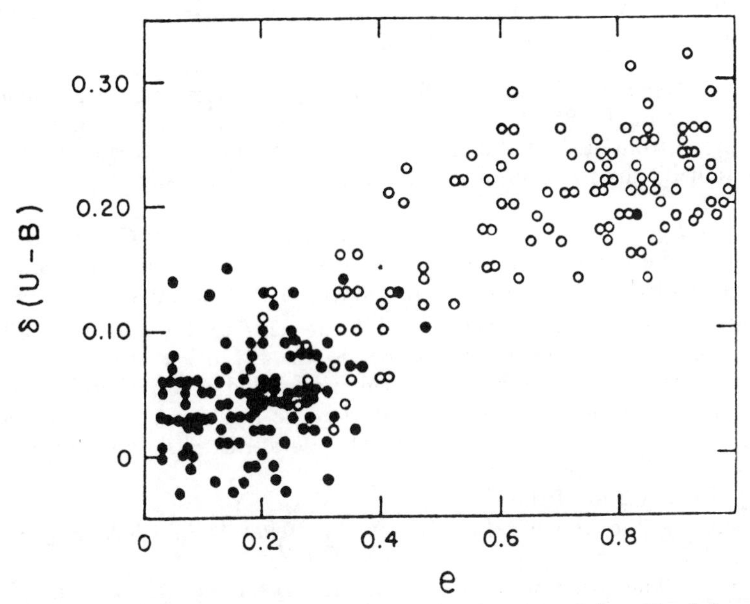

Figure 1 - The correlation between orbital eccentricity of subdwarf stars and their ultraviolet excesses, $\delta(U-B)$. Taken from Eggen, Lynden-Bell and Sandage (1962).

in the halo from those in the disk by their *distributions, their kinematics, their ages* or *their chemical compositions.*

2. A DEFINITION OF THE STELLAR HALO

2.1 How to tell halo stars from disk stars

Whatever may be the definition of the stellar halo or the stellar disk, there will certainly be some stars whose status as disk or halo is ambiguous. The purpose of the following discussion is to show that there are ways to flag at least some stars as being unambiguously halo stars, based on the preceding historical definitions for the halo. Briefly, the following definitions will separate stars that cannot belong to any other population:

- *Kinematical Definition* - Stars with velocities relative to the local standard of rest greater than 63 km sec^{-1} are halo stars. As mentioned above, the value of 63 km sec^{-1} was chosen by Oort(1922) to separate the two star streams; while the specific value is bit arbitrary, it is a reasonable limiting velocity for the purpose of defining halo stars.

- *Spatial Definition* - Stars more than 1000 pc from the galactic plane are halo stars. Since the scale height of the old thin disk is about 300 parsecs (Gilmore and Reid, 1983), at distances as large as

1000 pc from the plane, there should be a completely insignificant number of disk stars, although there will be some thick disk stars.

• *Chemical Definition* - Stars with [Fe/H] less than -1 are halo stars. Zinn (1985) has used this value to separate halo globular clusters from disk globulars; there is something of a gap in the metallicity distribution of globular clusters at that value.

• *Age Definition* - Stars with ages greater than 10 Gyr are halo stars. The oldest disk clusters are less than 10 Gyr in age (Janes, 1988) and the youngest globular clusters are significantly older.

These are exclusionary definitions, in the sense that a star fitting all these criteria could not be considered to belong to any other stellar population. However, some "true" halo stars will be excluded by these definitions. Furthermore, This definition of the halo for the most part *excludes* the thick disk component of the galaxy, including the metal-rich, disk globular clusters.

2.2 Global Properties of the Halo

Assuming the above definition for the stellar halo, what are its global properties?

• *Size* - The primary indicators of the size of the halo are the globular clusters, some of which are seen to distances as large as 100 kpc from the galactic center (for example, NGC 2419, Pal 3 or Pal 4) and distant RR Lyraes which are observed as far as 40 kpc from the galactic center (Saha, 1985). The stellar halo must be considered to be as much as 100 kpc in radius, although it is worth noting that the Large and Small Magellanic clouds are only about 50 and 60 kpc, respectively, from the galactic center (Feast, 1991).

• *Mass* - There are about 180 globular clusters in the halo (see sect. 6.1), and assuming an average globular cluster mass of 2×10^5 $M_\odot$ (Mihalas and Binney, 1981), the total mass in globulars is about $2 \times 10^7 M_\odot$. The local mass density of halo stars is about 100 times that of the globulars (Woltjer, 1975), from which one infers a halo mass of $2 \times 10^9 M_\odot$. Bahcall and Soniera (1980) derived a halo mass of $1.4 \times 10^9 M_\odot$ from a detailed model of the galaxy.

• *Luminosity* - The typical mass-to-luminosity ratio in globular clusters is about 1.5 (Webbink, 1985), leading to a halo luminosity of $2 - 3 \times 10^9 L_\odot$. When compared to the total galactic luminosity of $1.2 \times 10^{10} L_\odot$ (Bahcall and Soniera 1980), the halo makes an insignificant contribution to the galactic luminosity.

• *Shape* - The metal-poor globular clusters are distributed in an ellipsoid with an axial ratio of nearly 1:1, but Oort and Plaut (1975), find a ratio of 0.8 and Gilmore, *et al.* (1989) show evidence for a flattened halo with a ratio of about 0.6. The globulars may not be the appropriate tracer of the shape of the halo, simply because there are not enough of them.

• *Composition* - Applying the restriction that the metallicity, [Fe/H], of halo objects must be less than -1, the mean metallicity of the halo globular clusters is about [Fe/H] = -1.5, or Z = 0.0006 (Zinn, 1985). This amounts to only 1.2×10^6 $M_\odot$ of heavy elements in the entire halo, using the above estimate for the total mass of the halo.

If 0.5 $M_\odot$ of heavy elements is produced in a SN explosion (Arnett, 1978), then at the current rate of supernova explosions ($\sim$ 1 per 30 years), the entire heavy element content of the halo could have been generated in less than 100 million years. In other words, it is unlikely that the formation of the halo was marked by a great burst of supernova explosions.

3. COMPONENTS OF THE HALO

Although the primary focus of these lectures is the stellar halo, it is necessary to mention some of the other, non-stellar components of the halo.

3.1 The Globular Clusters

The 140 or so known globulars are the most obvious objects in the halo. The common assumption is that the globular cluster stars are representative of the general halo stellar population. The globular clusters will be discussed in detail in section 6.

3.2 Satellite galaxies

There are nine known galaxies within 250 kpc of the center of the galaxy (table 1). It is unknown whether they are gravitationally bound to the galaxy, or what their relationship to the halo is, but they are unquestionably intermingled with the other components of the halo. In particular, there are at least 10 globular clusters more distant than the Large Magellanic Cloud.

Table 1 - Satellite Galaxies in the Galactic Halo

Galaxy	R_{GC}	Galaxy	R_{GC}
LMC	50 kpc	Carina	94 kpc
SMC	60	Fornax	147
Ursa Minor	72	Leo I	213
Draco	75	Leo.II	234
Sculptor	78		

3.3 The gaseous component

There is a measurable amount of gas in the halo. Most of it is in the form of a hot (10^5K) medium, but there are cool neutral clouds as well. The gas in the halo may consist of material injected into the halo by supernova explosions (galactic fountains) or it may be gas falling into the galaxy from extragalactic sources or perhaps the Magellanic Stream. (See York, 1982, for a review of gas in the halo.)

3.4 The radiation field in the halo

Radiation *produced* in the galactic halo ranges from 21-cm radiation from cool clouds through optical and uv stellar radiation to far uv line emission from highly ionized halo gas. The halo is of course also bathed in radiation from the disk, including energetic photons from supernovae and hot disk stars. There may also be a significant far-uv and x-ray flux from extragalactic sources, such as quasars and active galactic nuclei. The radiation field has a negligible impact on the stellar population of the halo, but it is of course critical for understanding the gaseous component (York, 1982).

3.5 The problem of dark matter

This is a really BIG problem - as much as 90% of the mass of the galaxy is undetectable by any means other than gravitation, so the dark halo completely dominates the halo dynamics. Because it is so elusive and mysterious, the dark matter will be ignored in this discussion of the stellar halo.

4. STELLAR SURVEYS

Until recently, the study of galactic structure has mostly meant counting stars, and even now, surveys of stars along selected lines of sight can be used with great effectiveness in probing stellar populations. There are far too many stars to count them all, so one must choose particular lines of sight, or particular types of stars which most directly characterize the population of interest. The most primitive form of stellar survey is a simple count of the number of stars in a given direction, assuming all the stars are the same luminosity. Under this assumption, the number of stars is a measure of the distance to the edge of a uniformly populated galaxy.

4.1 Star Counts

We can be pretty sure that in general, both the assumption of uniform luminosity and uniform density of stars are incorrect. There will always be some sort of distribution of stars in space and with luminosity. The observed distribution of stars with apparent magnitude in solid angle ω is given by the star count equation,

$$A(m) = \omega \int_0^\infty \phi[m + 5 - 5logr - a(r)]D(r)r^2dr$$

where the luminosity function, summed over all spectral types, S, is given by

$$\phi(M) = \sum_S \phi(M, S)$$

the apparent distance modulus, including the interstellar absorption, $a(r)$, is

$$M = m + 5 - 5logr - a(r)$$

and $D(r)$ is the stellar density function.

This equation and its many applications are discussed more fully elsewhere, but I want to stress one point here that is often omitted. The solution of the star count equation, which is a convolution of the luminosity and density functions, is, like most convolution problems, not well-defined computationally. It is only effective at all if the analysis is based on *lots* of stars. We would also like to choose stars with a limited range of absolute magnitude, and if we could actually isolate stars of fixed luminosity, star count analyses and stellar population studies would be made much simpler. There are, of course, stellar types which more or less approximate this state. Thus, to study the stellar distribution in the halo, the simplest group of stars to deal with are the horizontal branch stars, especially the RR Lyraes. Red giant stars, being not terribly uniform in luminosity, present some difficulty, as do the subdwarfs, unless one can isolate an extremely limited range of spectral type, or at least account for the range in spectral types in some way.

The difficulty of the star count problem can be illustrated by considering the color-magnitude diagrams of typical stars. Figure 2 is a schematic composite color-magnitude diagram of various populations of stars, showing the zero-age main sequence (ZAMS) from Schmidt-Kaler (1982), together with a typical old open cluster color-magnitude diagram, M67 (Montgomery, *et al.*, 1993), a typical metal-poor globular cluster, M92 (Stetson and Harris, 1988) and a metal-rich globular, 47 Tuc (Hesser, *et al.*, 1987). These three clusters represent the range of properties of the old stellar populations of the galaxy; the vast majority of stars in the galaxy lie within those three sequences.

One must always keep this diagram in mind when doing star counts, or for that matter, almost anything related to stellar statistics. Thus, it is immediately obvious that by restricting the analysis to just the bluer stars, an old stellar sample (for example, along a line of sight away from the plane into the galactic halo) would consist of blue horizontal branch stars along with white dwarfs and quasars. The effects of the latter two groups on a blue star sample would be relatively easy to model. At the red end, the contributions of giants and dwarfs are also rather easy to model. However, at intermediate colors, the situation is much more complex, unless some way is found to separate distinct samples of stars by some other criterion than color.

Figure 3 illustrates the problem of sorting something out of a star count analysis. The distributions of stars in figure 3 can be generated by taking the color-magnitude diagram of a cluster like M92 and smearing it vertically, corresponding to a stellar population spread out over a range of distance. There is a characteristic "blue edge" to the distribution representing the main sequence turnoff region of old clusters. This is typically seen in the color-magnitude diagrams of stellar samples of old field stars. The temptation is to match the color of the blue edge with the turnoff of say, 47 Tuc, but caution is needed. As can be seen in figure 2, the main sequence turnoff of M67 is virtually identical in color to that of 47 Tuc, yet the latter is very old and somewhat metal-poor, whereas the former is a solar composition cluster, moderately old for a disk population, but still less than half the age of 47 Tuc.

4.2 Modern star count studies

The past few years have seen a rejuvenation of "classical" star count studies. Among the more important of the modern surveys was the work by Gilmore (see Gilmore, *et al.*, 1989) which led, among other things, to the discovery of the thick disk. Even grander surveys are on the way, especially here in Minnesota, home of APS machine with its capability for doing *real* star counts. Because the APS machine is able to count millions of stars, a detailed modelling will be possible of both the stellar mass function and the distribution function. The APS machine will be particularly useful for studies of the halo, and we can expect to hear more about this project as the survey becomes available.

Some recent studies directed toward individual classes of stars include the following:

The horizontal branch - In spite of the fact that the blue horizontal branch stars are relatively easy to isolate from other old stellar populations, this simple

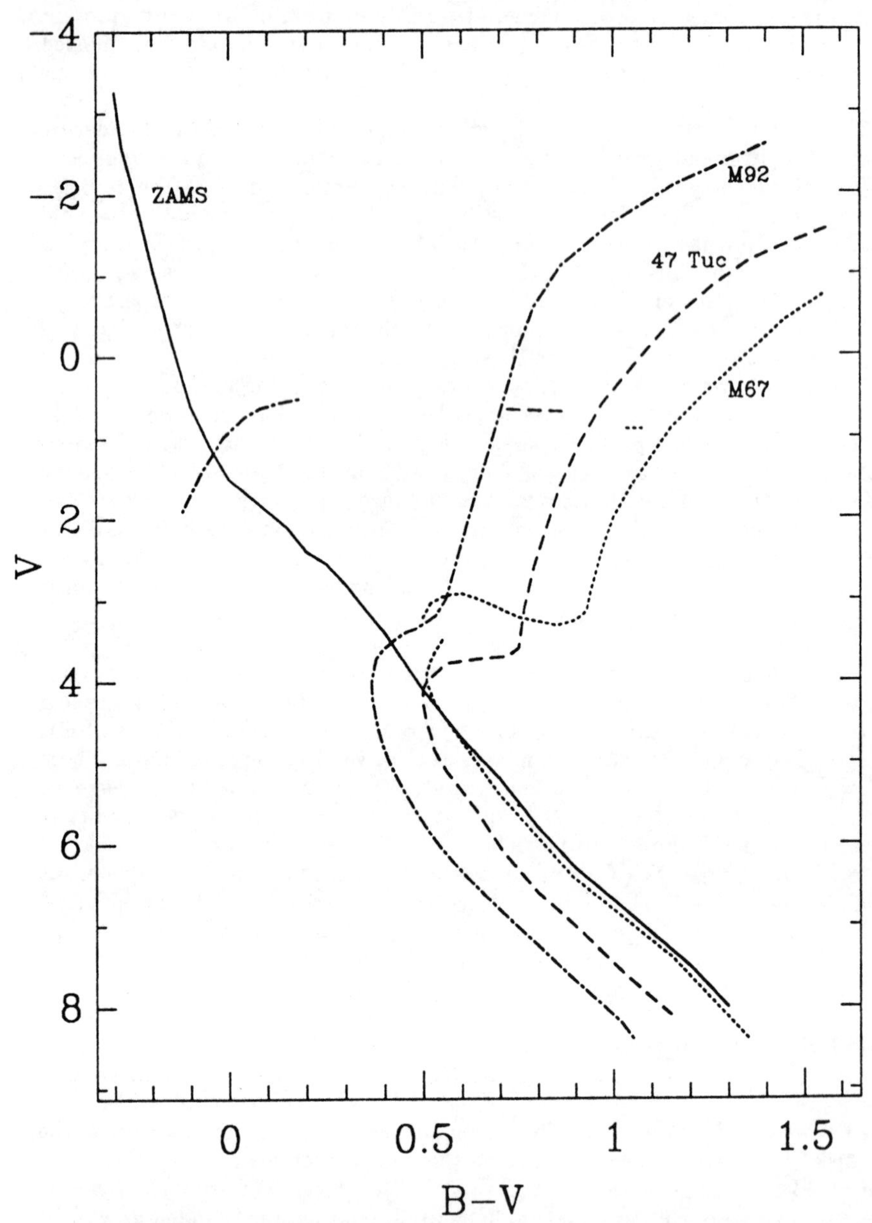

Figure 2 - Composite color-magnitude diagrams of the zero-age main sequence (Schmidt-Kaler, 1982), the old open cluster, M67 (Montgomery, *et al.*, 1993) the metal-poor globular cluster M92, (Stetson and Harris, 1988) and the metal-rich globular, 47 Tuc (Hesser, *et al.*, 1987).

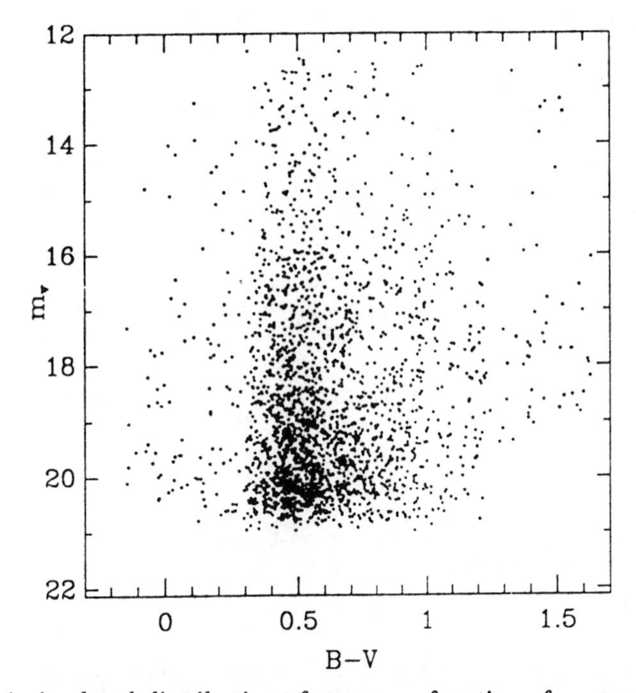

Figure 3 - A simulated distribution of stars as a function of apparent magnitude and color, as might be observed towards the galactic poles. This distribution was obtained by taking luminosity functions derived from M92 and 47 Tuc as shown in figure 2, and spreading them out in distance along a hypothetical line of sight. Note the apparent "edge" to the distribution near B-V = 0.4, corresponding to the M92 turnoff color.

fact has not been seriously exploited until recently. Preston, et al (1991) have studied the galactic distribution of several thousand blue horizontal branch stars. Among other results, they showed that the color of blue horizontal branch stars increases outward in the galaxy, just as the horizontal branches of globular clusters become more red in the outer galaxy. This is the "second parameter" effect (see section 6.4), which Preston, *et al.* take as a sign of an age gradient.

Rose (1985) and Norris and Green (1989) have debated whether the *red* horizontal branch stars that appear to be part of the thick disk are stars like those in the metal-rich globular clusters such as 47 Tuc, or whether they resemble instead the red giant "clump" stars found in old, somewhat metal-poor open clusters. Rose and Agostinho (1991) demonstrated that the red horizontal branch stars do indeed belong to the thick disk population, which is consistent with the idea that the thick disk is a moderately metal-rich population.

RR Lyrae stars - The RR Lyraes represent almost the ideal target for probing the character of the halo. They are found throughout the halo, the thick disk and the central bulge; their variability makes their identification unambiguous; they are moderately common; and most importantly, they have a very limited range in luminosity (M_V = 0.5 to 1.0), a luminosity that is now rather well determined. Some modern surveys of RR Lyrae stars include those by Kinman Wirtanen and Janes (1966), Plaut (1970), Kinman, *et al.* (1982) and Saha (1985). The net result of all these surveys is to confirm the $R^{-3.5}$ nature of the density of the galactic halo. Further, the Saha survey, the deepest of them, shows distinct signs of an "edge" to the halo at roughly 40 kpc from the galactic center.

The absolute magnitudes of the RR Lyrae stars are now rather well-defined. Several groups have recently applied the Baade-Wesselink method for determining the radius of an RR Lyrae from a comparison of its changes in radius as measured by radial velocities with the ratio of the radius that can be found from the magnitude differences at various points in its light curve. Liu and Janes (1991) used the surface brightness variation of the Baade-Wesselink method to derive the absolute magnitudes of 13 bright field RR Lyraes, plus 4 RR Lyraes in the globular cluster, M4. By adding data on 9 additional RR Lyraes from other sources, we were able to derive a relationship between the absolute magnitude of a star and its metallicity.

$$< M_V >= 0.20(\pm 0.07)[Fe/H] + 1.06(\pm 0.13)$$

Ultimately a more important relation connects the absolute K magnitude to the log of the period:

$$< M_K >= -2.44(\pm 0.24) logP - 0.90(\pm 0.13)$$

There are several potential advantages to using this relation. First, at the wavelength of the K-band, the effects of interstellar absorption are only about one-tenth as large as in the visual; the amplitude of the RR Lyrae variations is only a couple of tenths of a magnitude, so with just a few measures, a good approximation to the mean apparent magnitude can be found; and the period can be found with more than adequate precision from classical techniques.

Red Giant Surveys - Hartkopf and Yoss (1982) derived absolute magnitudes and compositions for red giant stars in the direction of the galactic poles and were able to probe stars out to several kpc from the galactic plane. The more distant halo stars in their sample are metal-poor, and the ones close to the plane have nearly solar composition; a plot of their data could be used to argue for a gradient in abundance decreasing steadily outward into the halo, which is the interpretation they take. This is also consistent with the ELS hypothesis for the collapse of the galactic halo. Norris and Green (1989) on the other hand, from a somewhat similar set of data argue that there could be a continuous distribution of disk populations, ranging from thick to thin disk with the scale height of each population being correlated with metallicity. Norris and Green contend that the halo and the disk are distinct from one another, but otherwise the differences between their conclusions and those of Hartkpof and Yoss are mostly semantic. One of Norris and Green's most important conclusions is the claim that there is no distinct thick disk population *per se*.

4.3 Kinematic and Abundance Studies

Luminous stars such as the red giants or RR Lyrae stars can be observed *in situ* throughout the halo, but the subdwarfs, the Population II main sequence stars, cannot be seen at large distances, although they are far more numerous. It is necessary instead to study local samples and infer properties of the halo from their kinematics and abundances. This leads to the problem of identifying the subdwarfs from the great mass of other types of stars. This can be done either through proper motion studies (kinematic selection) or through photometric and spectroscopic studies (abundance selection). In either case, there are inherent selection effects, and some *a priori* judgements have to be made about the character of the halo. This is a fundamental problem, one that has led to strong disagreements.

The ELS investigation of subdwarfs mentioned in the introduction is a classic study in the dangers of selection effects. They based their conclusions on the relationship between abundances and orbital eccentricity. In their sample, the metal-poor stars have large eccentricities, that is to say, they are moving on primarily radial orbits. However, their sample was derived from proper motion catalogues by selecting stars with the largest velocities. Thus, if there are metal-poor stars with low eccentricities, such stars would not have been included in their sample. Their central conclusion, that the metal-poor stars were formed in a collapsing system is based on the observation that they all have similar *radial* orbits. If a population of metal-poor stars were found with low eccentricities, their argument would be severely compromised. In fact, Norris, *et al.* (1985) have found just such stars which would lie in the upper left corner of figure 1.

Still, Sandage and Fouts (1987) argue on the basis of their analysis of 1125 high proper motion stars that the galaxy collapsed as a "continuous, coherent process rather than by mergers of disparate parts, each with their own enrichment history." In contrast, Norris and Ryan (1989) and Carney, Latham and Laird (1990) attempt to show that there is a distinct separation between the kinematics and abundances of disk and halo stars. The Carney *et al.* data, reproduced in figure 4, constitutes perhaps the strongest argument for a distinct halo.

Sandage and Fouts, Norris and Ryan and Carney *et al.* all addressed the issue of the thick disk, and whether there is a continuous gradation between halo stars and thin disk stars. Sandage and Fouts argued for a continuous sequence, somewhat like the ELS concept, whereas Carney *et al.* and Norris and Ryan concluded that there is a distinct non-rotating halo surrounding an independent rapidly rotating disk, but that separately within each population there is a range of parameters. Carney *et al.*, showed that the reason for the difference of opinion lies in the details of the presentation of the data, and by re-binning the Sandage and Fouts data, they could make it agree with their results.

5. DYNAMICS OF THE HALO

The mass of the stellar halo is much smaller than that of other components of the galaxy, so will have little dynamical impact on the rest of the system. However, the halo stars make good test particles for probing the rotation of the galaxy and the overall mass distribution in the galaxy.

5.1 Galactic rotation

In the "Keplerian" case, assuming a central point mass, we have the following standard relations:

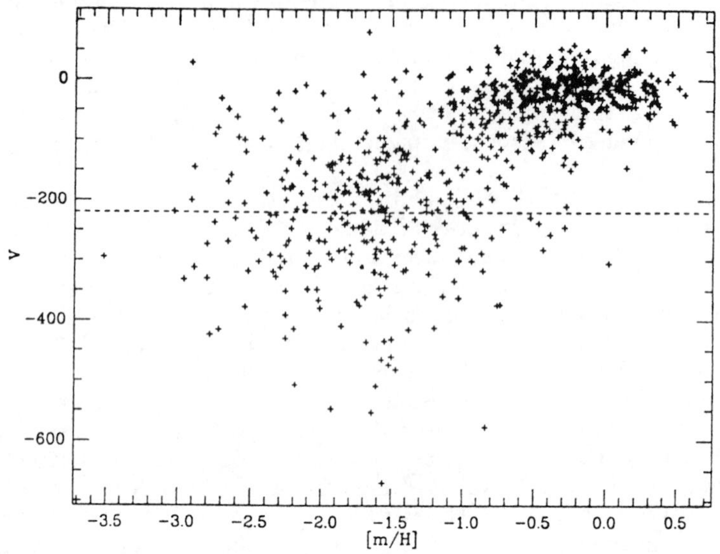

Figure 4 - Velocities in the V direction (directed towards galactic rotation, with respect to the local standard of rest) vs metallicity for 740 subdwarf stars with proper motions. From Carney, *et al.* (1990).

$$\phi(r) = -\frac{GM}{r}; \quad V_c(r) = (\frac{GM}{r})^{1/2}; \quad V_{esc}(r) = (\frac{2GM}{r})^{1/2}$$

At the solar position, $V_c(R_o) = 220\ km\ sec^{-1}$ so we would expect that $V_{esc}(R_o) = 308\ km\ sec^{-1}$. Oort's (1922) estimate of the escape velocity from the galaxy of 63 $km\ sec^{-1}$ relative to local standard of rest would lead to $V_{esc}(R_o) = 282\ km\ sec^{-1}$, a value in good agreement with the Keplerian estimate for the escape velocity from the solar neighborhood.

However, there are subdwarf stars in the solar vicinity known with velocities so large that if they are bound to the galaxy, then the local escape velocity must be in excess of 500 $km\ sec^{-1}$ (Carney, Latham and Laird, 1988). Such a large value cannot be reconciled with the bulk of the mass being interior to the Sun's position in the galaxy. These stars all have well-determined distances, proper motions and radial velocities, so it would be difficult to explain away the observations. Furthermore, the rotational velocity in the outer parts of our galaxy, as in many other galaxies, is known to be constant. Beyond the solar position, the rotational velocity of our galaxy is more less constant at just over 200 $km\ sec^{-1}$. Dynamically, the outer parts of our galaxy, especially the halo, cannot be dominated by a centrally concentrated mass distribution.

To get a flat rotation curve, we need:

$$V_c^2 = \frac{GM(r)}{r} = Constant$$

This implies that $M(r) \propto r$. But since

$$M(r) = \int_0^r 4\pi r^2 \rho dr$$

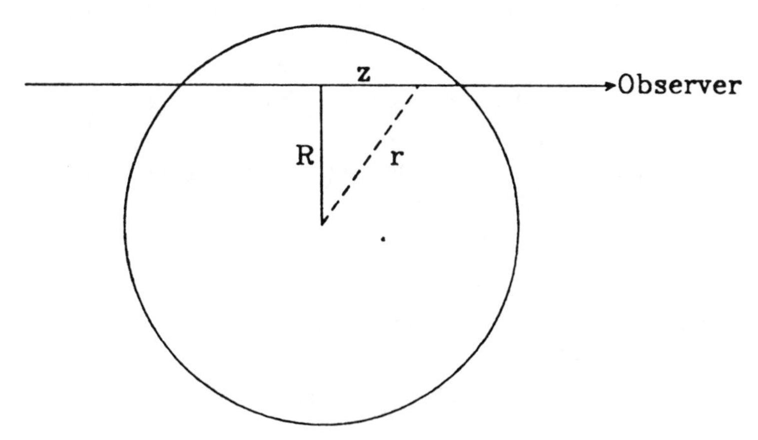

Figure 5 - The integrated luminosity, I(R) at a projected radius, R from the center of a galaxy is the integral of the density $\rho(r)$ along the line of sight z, assuming some mass-to-luminosity ratio.

then the constant rotation curve implies that the density of the galaxy is proportional to the radius squared. The combination of a large escape velocity and flat rotation curve almost *demands* the existence of large amounts of unseen matter.

It is possible to show (Binney and Tremaine, 1987) that in this situation, we can say

$$V_{esc}^2 = 2V_c^2[1 + ln(\frac{R_*}{r})] \quad for\ r < R_*$$

and

$$V_{esc}^2 = \frac{2V_c^2 R_*}{r} \quad for\ r > R_*,$$

where R_* is the "edge" of the dark matter halo. If we could measure V_{esc} at various large values of r, we could estimate R_* and the total mass of the galaxy.

5.2 Density and Luminosity Distribution of the Halo

A primary goal of star count studies is to determine the size and shape of the galaxy. This information in turn can be used first to compare with other galaxies and second to attempt to derive dynamical models for the galaxy. The problem is to find the density distribution in the galaxy.

In any galaxy, there exists some potential, $\phi(r)$, which implies a density distribution $\rho(r)$. We could just assume $\phi(r)$ based on first principles, or for our galaxy at least, we can measure $\rho(r)$ directly from star counts or kinematics. In external systems, we observe only a projected luminosity $I(R)$, as shown in figure 5. But I(R) is the integral of $\rho(r)$ along the line z, so to find $\rho(r)$, a deconvolution of the integral must be done, while making some sort of assumption about the mass-to-luminosity ratio, M/L.

A number of simple empirical relations have been derived over the years to describe density and luminosity profiles of spheroidal systems. See Binney and Tremain (1987) for a complete description. Some of the more important distributions include the *de Vaucouleurs law,*

$$I(R) = I(0)exp(-kR^{1/4}) = I_e exp\{-7.67[(\frac{R}{R_e})^{1/4} - 1]\},$$

where R_e is the effective radius, containing half the luminosity and I_e is the luminosity at R_e; the *modified Hubble profile,*

$$\rho(r) = \rho_o[1 + (\frac{r}{a})^2]^{-3/2},$$

where a is the core radius, which integrates to

$$I(r) = \frac{2I_o a}{1 + (\frac{R}{a})^2};$$

and a simple *power law,*

$$\rho(r) \propto (\frac{r_o}{r})^{\alpha},$$

which integrates to

$$I(R) \propto \frac{1}{R^{\alpha-1}}.$$

In the case of the power law, it can be shown that $M(r) \propto r^{(3-\alpha)}$, so for $\alpha < 3$, $M(r)$ diverges at large r. The density distribution of many elliptical galaxies and the halo of our galaxy are well-represented by an $r^{-3.5}$ relation.

6. GLOBULAR CLUSTERS

The key to understanding the halo is to understand the globular clusters. Not only are they the most prominent feature of the stellar halo, but we can also find their distances, velocities, compositions and especially ages rather precisely, at least compared to other objects.

6.1 Distributions of globular clusters

There are about 140 globular clusters known (Webbink, 1985). That is probably the vast majority of globulars in our galaxy, and Harris (1976) has estimated that the total population is of order 160 - 200. Thus, the current sample is about all we will ever have.

The sizes of globular clusters can be specified in two ways, by the core radius and by the tidal radius; the former ranges in size from 0.1 pc. to 25 pc. and the latter from 10 pc. to more than 100 pc. (Webbink, 1985). Quite obviously, there is a considerable range in the structural parameters of globular clusters – it is important to note that the largest clusters are not necessarily the most massive ones.

There are two distinct populations of globulars. The phrase "disk globulars' has been in use since at least the 1940's (Mayall, 1946; Kinman, 1959), but Zinn (1985) showed that there is a distinct break between the distributions of the globular clusters with [Fe/H] < -1.0 and those with [Fe/H] > -1.0. The halo globulars are distributed almost spherically around the galactic center, whereas the disk globulars are distributed in a highly flattened spheroid. Using the above metallicity criterion, Zinn finds 87 halo clusters and 28 disk clusters; the remaining clusters have unknown metallicities. The scale height of the disk globulars is greater than about 500 pc.

6.2 The Ages of Globular Clusters

The morphology of a globular cluster color-magnitude diagram depends on the cluster's age, its composition, its distance and the interstellar reddening to the cluster. The problem is to distinguish between old clusters affected in various degrees by these effects. The color-magnitude diagrams of old systems all look roughly the same, and estimating the relative contributions of the four parameters usually requires more information than just the basic photometry.

Nevertheless, we think we have a good idea of the ages of the globular clusters. Sandage (1953) deduced an age of 5 Gyr for the cluster M3, and Hoyle and Schwarschild (1955) applied early stellar models to derive an age of 6 Gyr. A few years later, Sandage (1962) derived an age of 22 Gyr for M3 using Hoyle's (1959) models. Demarque and Larson (1964) found an age for M3 between 17 and 25 Gyr. Since that time, there has been no systematic trend in the estimated ages of globular clusters. Table 2 shows calculated ages for M3; although there is a substantial range in age estimates, most of the differences are the consequence of different opinions as to the composition of the cluster. New generations of stellar evolution models and new, more precise observations have not changed the basic result.

Table 2 - Age estimates for the globular cluster, M3

Sandage (1953)	5 Gyr
Hoyle and Schwarschield (1955)	6
Sandage (1962)	22
Demarque and Larson (1964)	17 - 25
Iben and Rood (1970)	13 - 14.5
Sandage (1970)	10 - 12
Sandage and Katem (1982)	17
VandenBerg (1983)	18
Gratton (1985)	13.3 - 16.4
Paez, *et al.* (1990)	16 - 19
Sarajedini and King (1989)	14.6 - 17.6

In recent years, Population II models have been computed by a number of groups in several different countries. Some of the more recent models by these groups include Alongi, *et al.* (1990); Bergbusch and VandenBerg (1992); Castellani, *et al.* (1991); and Green, *et al.* (1987). In addition, there are a great many other people investigating various aspects of stellar evolution theory. This diversity, and competition among the various groups has helped to ensure that all conceivable aspects of the theory should be probed. Nevertheless, there are some features in common to all the calculations, and problems which have not yet been solved. So there are still some residual uncertainties about globular cluster ages, both their absolute ages, and their relative ages.

6.3 Uncertainties in Globular Cluster Ages

The uncertainties in the cluster ages can be divided into two categories, those resulting from uncertainties in the theory, and those resulting from uncertainties in the observations.

Model uncertainties – As mentioned above, the results of stellar evolution models have remained more or less unchanged in over 20 years. In that time, there have been a number of large and small fixes to opacities, energy generation, numerical techniques and other areas, but these fixes have not generally had much effect on the age estimates. Typically, corrections in one area have lead to compensating changes in the stellar parameters, so that the overall models remain about the same. In the end, the question boils down to the amount of nuclear fuel available (which depends on the star's mass) and the rate at which the fuel is burned (that is, the star's luminosity). Various complications that could be considered include rotation and helium diffusion. Some of these problems have recently been probed by, among others, Pinsonneault, *et al.* (1991) and Deliyannis and Demarque (1991), and the general agreement is that they are probably of minor importance to the evolution of population II stars.

A much more difficult problem is the interface between the theory and the observations. From the theory, one derives stellar effective temperatures and luminosities and from the observations, broad-band photometric color indices and apparent magnitudes are measured. The problem is to find a way to go between the two. Usually, the theoretical values are transformed into the observational plane, which requires a knowledge of bolometric corrections, detailed stellar atmospheres, and accurate instrumental response functions. Significant uncertainties exist in all of these.

The major unresolved problem on the theoretical side is the matter of turbulence. One might argue in fact that this is one of the great unsolved problems in all of physics; after all if we cannot model the atmosphere of the Earth and predict the weather, how can we expect to understand the turbulent material in the outer layers of a cool star? The present "mixing-length theory" is no theory at all, but an ad-hoc recipe for describing the convective energy flux through a star. Embedded in the theory is the infamous mixing-length parameter, α, the ratio of the distance a convective cell will travel before dissolution to the pressure scale height.

The present approach to matching theory with observation goes somewhat as follows:

1. Settle all the other physical parameters for a model
2. Compute a model for the Sun, that is to say, evolve a one solar-mass model of solar composition until the model star has an age of 4.5 Gyr.
3. Fiddle with α until the solar properties are duplicated in the model.

The interesting thing is that the approach seems to work rather well. Recent models of the Sun predict a value of α between 1.6 and 1.8, which is about what is derived from simple hydrodynamic computations. Furthermore, the derived depth of the convection zone in the Sun matches that derived from helioseismology measurements. Nevertheless, until someone does detailed, fully hydrodynamic models, there will be the residual uncertainty that perhaps under different conditions, the effects of convection will not be the same as in the Sun.

Observational uncertainties – Several possible areas of concern exist on the observational side. These include:

1. Overall photometric quality.

2. Interstellar reddening.
3. The distance scale.
4. Composition ("metallicity").
5. Helium abundance.
6. Oxygen, or α-product abundances.

The most problematic at the moment are probably numbers 4, 6, 3 and sometimes 2. The least problematic are 1, 5 and sometimes 2

The "Bottom Line" – In the determination of relative ages of two clusters of similar composition, it is now possible to achieve a precision of significantly better than one Gyr. For absolute ages, the uncertainties are still probably of the order of 4 Gyr. Nevertheless, to get the oldest globular clusters, such as M92, to an age of less than 13 Gyr, which seems to be required by most cosmological models, will probably require some sort of new physics.

6.4 Age differences among globular clusters

If something like the ELS scenario for the formation of the halo is correct, the globular clusters should all have the same age, within current uncertainties. At first glance, that appears to be the case. The major differences in the morphology of globular cluster color-magnitude diagrams are due to differences in metallicity. In particular, most metal-poor clusters have blue horizontal branches. However, some metal-poor clusters have red HB's, similar to the HB morphology of more metal-rich clusters. Almost all of these are in the outer halo. The question is, why?

This is the so-called globular cluster "second parameter problem", and is based on a single observational fact: The morphology of the horizontal branches of globular clusters in the outer halo differ from those in the inner halo. The horizontal branches of metal-poor clusters in the inner halo contain mostly blue stars, but the stars in similar clusters of the outer halo are often red instead. Possible solutions include differences in age, helium abundance, rotation, and anomalies in CNO abundances, and although a difference in age between the inner and outer clusters is often cited as the explanation, at the moment, there is no agreement on the physical causes of the phenomenon (see, *e.g.*, VandenBerg and Durrell, 1990).

There are definitely some clusters with ages different from others of the same composition. The most striking example is NGC 288 which must be several Gyr older than the otherwise identical cluster, NGC 362 (Sarajedini and Demarque, 1990). Ru 106 in the outer halo is extremely metal-poor, but substantially younger than clusters of similar composition in the inner halo (Buonanno, *et al.* (1990). The evidence is growing that star formation in the outer halo took place over an extended period of time.

7. DEVELOPMENT OF THE GALACTIC DISK

As described above, there is persuasive evidence that our galaxy did not collapse suddenly out of a simple, homogeneous gas cloud soon after the Big Bang. Instead it is likely that the formation of the galactic halo was a prolonged, somewhat chaotic affair, lasting several billion years (Larson, 1990). Not only are the oldest globular clusters of the order of 15 Gyrs in age or even older, depending on such possible complications such as variable [O/Fe] or Helium diffusion (Profitt and VandenBerg, 1991), but as mentioned above, we now know that some globular clusters are several billion years younger than others (Sarajedini and Demarque, 1990).

So how does the galactic disk fit in to the picture? In contrast to the halo, the oldest stars of the galactic disk are measured to be less than 10 billion years in age (Janes, 1988). Furthermore, recent observations, such as those by Carney, *et al.* (1990), leave one with the distinct impression that the disk and the halo are rather disconnected from one another. The Searle and Zinn (1978) hypothesis is that the halo formed from pre-galactic fragments of the size of globular clusters or perhaps an early generation of dwarf galaxies. If so, then the disk developed in a separate process which is left unexplained. The disk and halo cannot really be unrelated entities of course, and there must have been some sort of progression between the early formation of the halo and the apparently much later development of the disk. Zinn (1992) has recently argued that the *halo* globular clusters are themselves divided into two groups, the "Old Halo" which formed in the collapse that led eventually to the formation of the disk, and a "Younger Halo" consisting of fragments from disrupted satellite systems accreted by the galaxy.

A plausible extension of this hypothesis for the formation of the halo is that the disk developed gradually as material continued to rain onto the galaxy for a long period of time, perhaps even up to the present. The occasional impact of small galaxies or hydrogen clouds has kept the pot boiling, diluting the interstellar medium with metal-poor gas, and ejecting stars and clusters into the halo (see Larson, 1990; Tenorio-Tagle, *et al.*, 1987)

New hydrodynamic models for the development of the galaxy such as those by Katz and Gunn (1991) that take into account dark matter confirm this idea and show that the disk formed only gradually as material settled into the galaxy. Recent chemical evolution models by Sommer-Larsen and Yoshii (1990) also require a prolonged period of infalling proto-galactic material to form the disk.

We certainly do not fully understand the halo of our galaxy, either how it formed, or how and when the galactic disk developed out of the halo. But we must remember that since the halo stars date from that formation period, they are the only nearby artifacts of the early days of the universe. It should also be particularly evident to the readers of this section that the "dark matter problem", which has been so studiously avoided here, is really central to our eventual understanding of the formation and the evolution of the galaxy.

ACKNOWLEDGEMENTS

It has been a pleasure to visit the University of Minnesota, and I am grateful to Roberta Humphreys for inviting me to deliver these lectures. Some of my research mentioned in these pages has been supported in part by National Science Foundation grants AST-8818360 and AST-8915444.

REFERENCES

Alongi, M., Bertelli, G., Bressan, A. and Chiosi, C. 1990, in The formation and Evolution of Star Clusters, K. Janes, Ed., (San Francisco: Astronomical Society of the Pacific), p. 223.
Arnett, W.D. 1978, *Ap. J.*, 219, 1008.
Baade, W. 1944, *Ap. J.*, 100, 137.
Bahcall, J.N. and Soneira, R.M. 1980, *Ap. J. Suppl.*, 44, 73.
Bergbusch, P.A. and VandenBerg, D.A. 1992, *Ap. J. Suppl.*, 81, 163.
Binney, J. and Tremaine, S. 1987, Gâlactic Dynamics (Princeton: Princeton Univ. Press).
Buonanno, R., Buscema, G., Fusi Pecci, F., Richer, H.B. and Fahlman, G.G. 1990, *Astron. J.*, 100, 1811.
Carlberg, G., Dawson, P.C., Hsu, T. and VandenBerg, D.A. 1985, *Ap. J.*, 294, 674.
Carney, B.W., Latham, D.W. and Laird, J.B. 1988, *Astron. J.*, 96, 560.
Carney, B.W., Latham, D.W. and Laird, J.B. 1990, *Astron. J.*, 99, 572.
Castellani, V., Chieffi, A. and Pulone, L. 1991, *Ap. J. Suppl.*, 76, 911.
Deliyannis, C.P. and Demarque, P. 1991, *Ap. J.*, 379, 216.
Demarque, P. and Larson, R.B. 1964, *Astron. J.*, 69, 136.
Eggen, O.C., Lynden-Bell, D. and Sandage, A.R. 1962, *Ap. J.*, 136, 748 (ELS).
Feast, M.W. 1991, in The Magellanic Clouds, R. Haynes and D. Milne, Eds., (Dordrecht: Reidel), p.1.
Gilmore, G. and Reid, N. 1983, *M.N.R.A.S.*, 202, 1025.
Gilmore, G., Wyse, R.F.G. and Kuijken, K. 1989, *Ann. Rev. Astron. Astrophys.*, 27 555.
Gratton, R.G. 1985, *Astr. Ap.*, 147, 169.
Green, E.M. Demarque, P. and King, C.R. 1987, The Revised Yale Isochrones and Luminosity Functions (New Haven: Yale University Observatory).
Harris, W.E. 1976, *Astron. J.*, 81, 1085.
Hartkopf, W.I. and Yoss, K.M. 1982, *Astron. J.*, 87, 1679.
Hesser, J.E., Harris, W.E., VandenBerg, D.A., Allwright, J.W.B., Shott, P. and Stetson, P.B., 1987, *Pub. A.S.P.*, 99, 739.
Hoyle, F. 1959, *M.N.R.A.S.*, 119, 124.
Hoyle, F. and Schwarszchield, M. 1955, *Ap. J. Suppl.*, 2, 1.
Iben, I. and Rood, R.T. 1970,*Ap. J.*, 161, 587.
Janes, K.A. 1988, in Calibrating Stellar Ages, A.G.D. Philip, Ed., (Schenectady: L. Davis Press). p. 59.
Katz, N. and Gunn, J.E. 1991, *Ap. J.*, 377, 365.
Kinman, T.D. 1959, *M.N.R.A.S.*, 119, 538.
Kinman, T.D., Mahaffey, C.T. and Wirtanen, C.A. 1982, *Astron. J.*, 87, 314.
Kinman, T.D, Wirtanen, C.A. and Janes, K.A. 196, *Ap. J. Suppl.*, 13, 379.
Larson, R.B. 1990, *Pub. A.S.P.*, 102, 709.
Linblad, B. 1925, *Ap. J.*, 62, 191.
Liu, T. and Janes, K.A. 1991, in *The Formation and Evolution of Star Clusters*, K.A. Janes, ed., (San Francisco, ASP) p. 278.
Mihalas, D. and Binney, J. 1981, Galactic Astronomy, 2nd Ed., (San Francisco: Freeman), p. 122.
Montgomery, K., Marschall, L.A. and Janes, K.A. 1993, *Astron. J.*, in press.
Norris, J.E. and Green, E.M. 1989, *Ap. J.*, 337, 272.

Norris, J.E. and Ryan, S.G. 1989, *Ap. J.*, 340, 739.
O'Connell, D.J.K., ed. 1958, "Stellar Populations", (Amsterdam: North-Holland).
Oort, J.H. 1922, Bull. Astron. Inst. Netherland, 1, 133.
Oort, J.H. 1927, Bull. Astron. Inst. Netherland, 3, 275.
Oort, J. and Plaut, L. 1975, *Astr. Ap.*, 41, 71.
Paez, E., Straniero, O. and Martinez Roger, C. 1990, *Astr. Ap. Suppl.*, 84, 481.
Pinsonneault, M.H., Deliyannis, C.P. and Demarque, P. 1991, *Ap. J.*, 367, 239.
Plaut, L. 1965, *Astr. Ap.*, 8, 341.
Preston, G.W., Schectman, S.A. and Beers, T.C. 1991, *Ap. J.*, 375, 121.
Proffit, C.R. and VandenBerg, D.A. 1991, *Ap. J. Suppl.*, 77, 473.
Roman, N.G. 1952, *Ap. J.*, 116, 122.
Rose, J.A. 1985, *Astron. J.*, 90, 787.
Rose, J.A. and Agostinho, R. 1990, *Astron. J.*, 101, 950.
Saha, A. 1985, *Ap. J.*, 289, 310.
Sandage, A.R. 1953, *Astron. J.*, 58, 61.
Sandage, A.R. 1962, *Ap. J.*, 135, 349.
Sandage, A.R. 1970, *Ap. J.*, 162, 841.
Sandage, A.R. 1982, *Ap. J.*, 252, 553.
Sandage, A.R. and Fouts, G. 1987, *Astron. J.*, 93, 74.
Sarajedini, A. and Demarque, P. 1990, *Ap. J.*, 465, 219.
Sarajedini, A. and King, C.R. 1989, *Astron. J.*, 98, 1624.
Schmidt-Kaler, T. 1982, in Landolt-Bornstein VI, Vol 2b, (Berlin: Springer-Verlag), p. 1.
Searle, L. and Zinn, R. 1978, *Ap. J.*, 225, 357.
Shapley, H. 1918a, *Ap. J.*, 48, 154.
Shapley, H. 1918b, *Pub. A.S.P.*, 30, 42.
Sommer-Larsen, J. and Yoshii, Y. 1990, *M.N.R.A.S.*, 243, 468.
Stetson, P.B. and Harris, W.E. 1988, *Astron. J.*, 96, 909.
Tenorio-Tagle, G., Franco, J., Bodenheimer, P. and Rozycka, M. 1987, *Astr. Ap.*, 179, 219.
VandenBerg, D.A. 1983, *Ap. J. Suppl.*, 51, 29.
VandenBerg, D.A. and Durrell, P.R. 1990, *Astron. J.*, 99, 221.
Webbink, R.F., 1985, in Dynamics of Star Clusters, J. Goodman and P. Hut, eds., (Dordrecht: Reidel), p. 541,
Woltjer, L. 1975, *Astr. Ap.*, 42, 109.
York, D.G. 1982, *Ann. Rev. Astr. Ap.*, 20, 221.
Zinn, R.J. 1985, *Ap. J.*, 293, 424.
Zinn, R.J. 1992, in The Globular Cluster-Galaxy Connection: Globular Clusters in the Context of their Parent Galaxies, G.H. Smith and J.P. Brodie, Eds., (San Francisco: Astronomical Society of the Pacific), in press.

The Minnesota Lectures on Clusters of Galaxies and Large-Scale Structure
ASP Conference Series, Vol. 39, 1993
Roberta M Humphreys (ed.)

GALAXY MODELS AND GALACTIC MODELS

GERARD GILMORE
Institute of Astronomy, Madingley Road, Cambridge, England

ABSTRACT Modelling of observational data is required so that one can extract parameters and functions, and their limits of validity, which both adequately retain the information content of the data and yet allow practical derivation of quantities of general astrophysical interest. Modelling galaxies from first principles is required to identify the most important physical processes, and to provide testable predictions to verify or refute the models. Examples of each of these types of modelling are given. The modelling of observations is illustrated by deriving the stellar initial mass function for low mass field stars. This process starts with a mix of parallax and photometric data for nearby single stars, and photometry for more distant stars. The distant stars are mostly unresolved binaries, and have a different distribution of chemical abundances, ages, and measuring errors than do the local data. All these effects are important in the modelling. Additionally, the space density of stars varies due to large and small scale structure of the Galaxy. Thus one must model the properties of stars and of Galactic structure at the same time to derive the stellar IMF. One of the more interesting applications of modelling the Galaxy is for comparison with models of galactic formation. Galaxy formation models and a practical astrometric test of the triaxiality of cold dark matter halo models are described to illustrate *a priori* modelling and its testing.

INTRODUCTION

There are two types of modelling of any astrophysical data. One, which may be termed "data reconstruction" seeks to identify a minimal but adequate set of functions and parameters which describe a data set. A simple common example is the use of a mean and gaussian dispersion to describe a set of measurements of stellar velocities. The second type, which may be termed *"a priori"* aims rather to identify the range of physical processes and boundary conditions which can have led to an observed distribution. An example is the use of velocity diffusion in kinematic phase space, based eventually on parameters describing the distribution of mass inhomogeneities in the Galaxy, to reproduce the approximately gaussian distribution of velocities of old stars near the Sun.

The general aims of all modelling are to provide convenient representations of data sets in such a way that one may learn or test some (astro)physics. For the specific case of interest here, modelling Galactic structural data, many levels of sophistication are possible. Some obvious relationships can be summarised as follows:

Type of Data		Model Function and Physics
Kinematics	$\Longleftrightarrow$	Dynamics
radial velocities proper motions		phase space distribution function spatial distributions $\Rightarrow$ gravitational potential dissipational history
Chemical Abundances	$\Longleftrightarrow$	Chemical Evolution
line strengths, photometry		star formation history ISM history $\Rightarrow$ stellar initial mass function gas flows, dissipation, SFR
Star Counts	$\Longleftrightarrow$	Galactic Structure
colour magnitude data surface brightness		spatial distribution function luminosity functions $\Rightarrow$ stellar IMF, binarism, dissipation, SFR, ...

Several of the important physical processes are relevant to each of the three physical examples noted above, emphasising the inter-relationships between all the apparently diverse types of data. In fact, the distribution of stars in the Galaxy in coordinate space, in kinematic space, and in chemical abundance space, are intimately related. A meaningful understanding of any sub-part of stellar phase space presupposes consideration of all other parts.

In the rest of this article we give one example of detailed modelling of observational data to derive reliable functional forms of physical significance – Galactic modelling, and one example of *a priori* modelling – galactic modelling. The detailed model is the modelling of stellar number-magnitude-colour data to derive the stellar luminosity function and its derivative the stellar mass function, and the spatial distribution of stars in the Milky Way. A more general description of the observational consequence of current galaxy formation models is followed by an example of an observational project which could help test current cold Dark Matter models.

MODELLING STAR COUNTS

The number of stars observed in a given range of apparent magnitude and colour is calculated from

$$
\begin{aligned}
N(m,\ \text{colour}) = \omega \int dV \quad & \cdot\ \Phi\left(M_{v_1}, M_{v_2}, \left[\frac{m}{H}\right], \left[\frac{\alpha}{H}\right], \tau, D, \ldots\right) \\
& \cdot\ D\,(F, M_{v_i}, \tau, \ldots) \\
& \cdot\ M_{v_i}\ (\text{colour}) \qquad (1)
\end{aligned}
$$

where the luminosity function Φ, the spatial density distribution D, and the colour-luminosity relation M_v (colour) all depend on

- $-M_{v_i}$ – stellar luminosity, which really depends on mass, age (τ), and chemical abundance ($[m/H]$, $[\alpha/H]$)
- $-[m/H]$ – the overall chemical abundance
- $-[\alpha/H]$ – the relative distribution of heavy elements
- $-\tau$ – age, particularly through stellar evolution and possibly through secular changes in other variables
- $-$distance $\bar{r}$ – through the age-location relation, and possibly through other systematics,

and so on.

An important point about this complex set of relationships is that most variables have ranges, so that both random (directly) and systematic (through the Malmquist effect) changes are inevitable.

Systematic corrections for systematic effects

An example of modelling data with a *systematic* change in a calibrating variable is given by the derivation of the spatial density profile perpendicular to the Galactic disk by Kuijken and Gilmore (1989). In this case the technique of photometric parallax was applied to a sample of K dwarfs. However, the systematic decrease in mean metal abundance with increasing distance leads to a systematic difference $\Delta M(z)$ between the correct distance and that derived using an assumed solar abundance, z_0

$$z = z_0 10^{-0.2\Delta M(z)} \tag{2}$$

This leads to a change in volume element

$$\frac{dV}{dV_0} = \frac{\omega z^2 dz}{\omega z_0^2 dz_0} = \frac{z^3 d\log z}{z_0^3 \log z_0} \tag{3}$$

$$= 10^{-0.6\Delta m}\left\{1 + \frac{0.2z}{\log e}\frac{d\Delta m}{dz}\right\}^{-1} \tag{4}$$

so that the density law $\nu(z)$ becomes

$$\nu(z) = 10^{0.6\Delta m}\left(1 + 0.4605\frac{d\Delta m}{dz}\right)\nu_0\left(z.10^{0.2\Delta m}\right). \tag{5}$$

Substitution of an adopted range of abundance gradients, through their systematic effect affects Δm on the photometric calibration, then provides the necessary correction.

Systematic Corrections for Random effects

Random errors produce systematic biases in truncated samples. The most common example of this is the Malmquist bias. The effect arises in a distance (or magnitude) limited sample where there is an intrinsic range in the distance calibration in use. A simple way to think of this effect is that, when the density of sources changes across a boundary, and each source has a random distance error associated with it, more sources will scatter from the higher source density region to the lower. This statistical osmosis is easily exemplified by writing equation (1) in distance modulus

$$A(m) = \omega \int D(\rho)\Phi(m + 5 - 5\log\rho)\rho^2 d\rho. \tag{6}$$

Following the application by Malmquist to star count studies, we model the luminosity function Φ as a sum of gaussians and residuals. This is a very good approximation if one restricts data to a narrow range of colour and luminosity class, as indeed is usually the practise. Then we have

$$\Phi = \frac{N(m)}{\sqrt{2\pi}\,\sigma_m} \exp\left\{-\frac{[M - \bar{M}(m)]^2}{2\sigma_m^2}\right\} + A_3\Phi''' + \ldots \tag{7}$$

and, defining $q = dA(m, r)$ so that $\int q dr = A(m)$,

$$\bar{M}(m) = \frac{\int M(m, r) q(m, r) dr}{\int q(m, r) dr} \tag{8a}$$

(where $\bar{M}(m)$ is the mean absolute magnitude in a magnitude-limited sample)

$$= \frac{\omega}{A} \int M(m, r)\Phi(m, r) D(r) r^2 dr \tag{8b}$$

$$= \frac{\omega}{A} \frac{Q_0}{\sqrt{2\pi}\,\sigma} \int (m + 5 - 5\log r) \exp\left\{\frac{[m + 5 - 5\log r - M_0]^2}{2\sigma^2}\right\} D.r^2 dr \tag{8c}$$

In terms of observed number counts, dA/d_M, and using M_0 as the mean absolute magnitude in a volume-limited sample,

$$\begin{aligned} \bar{M}(m) &= M_0 - \frac{\sigma^2}{A}\frac{dA}{dm} \\ &= M_0 - \frac{\sigma^2}{\log e}\frac{d\log A}{dm} \end{aligned} \tag{9}$$

An analogous derivation for the dispersion of a magnitude-limited sample is

$$\sigma_m^2 = \sigma^2 \left\{1 + \frac{\sigma^2}{\log e}\frac{d^2 \log A}{dm^2}\right\}. \tag{10}$$

It follows that increasing apparent number counts $dA/dm > 0$, lead to a bias such that a typical luminosity of a source in a luminosity-limited sample exceeds that of the parent population from which the sample is drawn, $\bar{M}(m) < M_0$.

It is important to remember that the amplitude and the sign of the Malmquist correction depend on both the luminosity function Φ and the apparent number counts $d(\log A)/dm$. Thus Malmquist corrections can have very different amplitudes and a different sign depending on the data set. Looking radially out in the Galactic disk can be very different than looking radially in.

A very similar effect, known as the Lutz-Kelker effect, in parallax work, arises when sample selection imposes a bias on the symmetry of the allowed error distributions, in this case because a negative parallax is forbidden. It is a salutary experience to overlay absolute magnitude-colour H-R diagrams for high quality and low quality parallax data, and observe the systematic shift in location of the main sequence.

We now give an example of how one models a real astronomical data set, in which both random and systematic effects must be understood. The example chosen is the derivation of the stellar luminosity function and its derivative, the stellar mass function, for low mass stars near the Sun, by Kroupa, Tout and Gilmore (1990, 1991, 1992).

MODELLING THE STELLAR LUMINOSITY FUNCTION.

There are two almost independent determinations of the stellar luminosity function (LF) for low mass stars. One is based on detailed analysis – trigonometric parallax, kinematics, high angular resolution imaging – of the complete sample of stars very near the Sun. The best studied data set is restricted to the sky north of $\delta = -20°$ and within 5.2 pc. The second LF is determined from photometric data for stars at larger distances – 50 pc to 100 pc being possible – with distances derived by photometric parallax.

Each of these LF's has advantages and disadvantages. The 5.2 pc LF is based on high precision data for single stars, since most binaries will have been resolved or detected. However, the sample is small, so that the statistical precision of the LF is poor. The more distant sample has very many stars, so that random errors are unimportant. In this sample however close binaries remain unresolved, so that LF is that of stellar systems, not of single stars. Additionally, it is essential that the photometric parallax method – essentially application of the absolute magnitude-colour relation in reverse – is corrected from its local calibration to be appropriate to the field sample of relevance. As well as this calibration, the various contributions to the scatter in the absolute magnitude-colour (CM) relation must be understood, so that appropriate Malmquist corrections can be applied.

The important parameters in this experiment are the distributions of ages, chemical abundances, binaries, and measuring errors, each of which can change systematically from local to distant samples, and each of which should be modelled. With this diversity a Monte Carlo approach is most effective.

Sources of Astrophysical Scatter

Those variables which need to be considered in the modelling include a mixture of things one must calibrate ("noise") and things one wants to learn ("signal").

Pre main-sequence evolution

Low-mass stars evolve onto the main sequence over a considerable period. During this time predictions of the luminosity are complicated by asymmetric dust shells and disks, rapid mass loss, chromospheric activity and reddening from residual molecular cloud material. Thus quantification of this effect is difficult. Fortunately, for present purposes, most stars at high latitudes are old enough to be free of such complexities. When modelling luminosity functions in young open clusters, however, the problems can be considerable.

Main-sequence stellar evolution
Stars evolve up and across the main-sequence during their lives. Evolutionary tracks are available for intermediate and high mass stars, and for abundances up to about solar. For lower masses and higher abundances some extrapolation is needed. Analogous to this of course one must adopt an age distribution. For high latitude field stars a uniform distribution is appropriate.

Chemical Abundances
Chemical abundances systematically affect the location of the main sequence. The correction for this is straightforward in principle and is detailed in equation 5. Complexities arise in quantifying the abundance effect at low luminosities, and in knowing the relevant abundance distribution ("metallicity gradient") to adopt. For low luminosities, where very few subdwarfs with good parallaxes are known for an empirical calibration, extrapolation of atmospheric models is required. For abundance ranges of relevance, some observational data are available. The most reliable method is however to investigate the allowed range of possibilities (Kuijken and Gilmore 1989), both in the mean and range.

Parallax Errors
Parallax errors in both the nearby LF sample and in the sample defining the colour-luminosity relation (if these differ) must be considered. Both the random and the resulting systematic bias need consideration, following equation 9 and Lutz and Kelker (1973).

Binary Stars
The number of unresolved binaries is a critical parameter which varies from sample to sample. The number will be least in the 5.2 pc sample, as it is both close (enhancing spatial resolution) and well studied. In the sample used to define the CM relation it will be greater, though many wide binaries will have been found and perhaps rejected from the sample, during parallax studies. For the field LF sample the spatial resolution of the mostly photographic surveys exceeds several arcsecs, so that the full binary properties of the LF are relevant. It is of course possible that there is a correlation between primary and secondary properties in the binary mass function, or a correlation between the incidence of binarism and primary mass. All these possibilities must be included in the range of models.

Galactic Structure
When comparing the space density of stars per unit volume of samples in the Galactic Plane and far from the Plane it is clearly necessary to allow for the systematic change in space density with distance, the density law. In fact, this exercise is in large part designed to determine this.

The effect of all these contributions to the error budget is to confuse a reliable comparison of the nearby single star luminosity function with the distant stellar system luminosity function. The apparent complexity of this process however illustrates an important general point. It is usually more reliable to know the allowed range of values of a function or a parameter than it is to know the "best" available determination.

In this case calculation proceeded using a range of parameters in a Monte Carlo process, as described by Kroupa *et al.* (1990, 1991, 1992). The first requirement is to check that reasonable ranges of parameter space are considered, by reproducing a well understood case. Here the most appropriate example is the "cosmic scatter" of $\sigma \sim 0.^m3$ about a mean colour-magnitude relation for well-studied trigonometric parallax stars. With that successfully achieved one can begin to do astrophysics.

The specific null hypothesis of relevance to be tested is: Is there a single stellar mass function, convolved with a realistic stellar binary distribution and a Galactic disk scale height, which can be converted through a consistent mass-luminosity relation to reproduce both the local single star luminosity function and the more distant stellar system luminosity function?

Following calibration of the modelling technique as described below, it is next necessary, prior to answering the question, to consider the mass-luminosity calibration.

Conversion of Luminosity to Mass

In the section above we have discussed conversion of apparent colour of a stellar system to the space density of single stars as a function of their intrinsic luminosity, the luminosity function. A more fundamental function is the space density of single stars as a function of their mass, the present day mass function (PDMF). For the low mass stars of interest here, where mass loss during that part of their main sequence lifetimes which has occurred to date is small, the mass function is a good approximation to the initial mass function (IMF) generated by the star formation process. For higher mass stars, where evolutionary corrections are large, derivation of the IMF from the PDMF is a complex exercise. [An excellent discussion of this more general problem is provided by Scalo (1986).]

The stellar luminosity function $\Phi(M_v)$ is directly related to the stellar mass function $\zeta(\mathcal{M})$ as

$$\Phi(M_v) = \frac{dN}{dM_v} = -\frac{dN}{d\mathcal{M}} \cdot \frac{d\mathcal{M}}{dM_v} = -\zeta \frac{d\mathcal{M}}{dM_v}. \tag{11}$$

The important point here is that the *gradient* of the mass-luminosity relation is the function which relates the LF to the IMF.

The existence of structure in the mass-luminosity relation thus generates structure in the LF, even for a smooth underlying IMF. This is illustrated in Figure 1, which shows both the LF for single stars within 5.2 pc of the sun (Figure 1a) and the observational constraints on the mass-luminosity relation. The most obvious features of the LF are a flat section near $M_v \sim +7$, and a maximum near $M_v \sim +12$. Were the mass-luminosity relation to have a nearly constant gradient then these features would be reproduced in the IMF, and would represent properties of the physics of star formation. Unfortunately, they are artefacts of molecular physics, in that the mass luminosity relation has structure at these luminosities. This is illustrated in Figure 2, which shows the *gradient* of the mass-luminosity relation. The important feature of Figure 2 is that features in the mass-luminosity relation correspond to those in the luminosity function. This simple correspondence means that the stellar mass

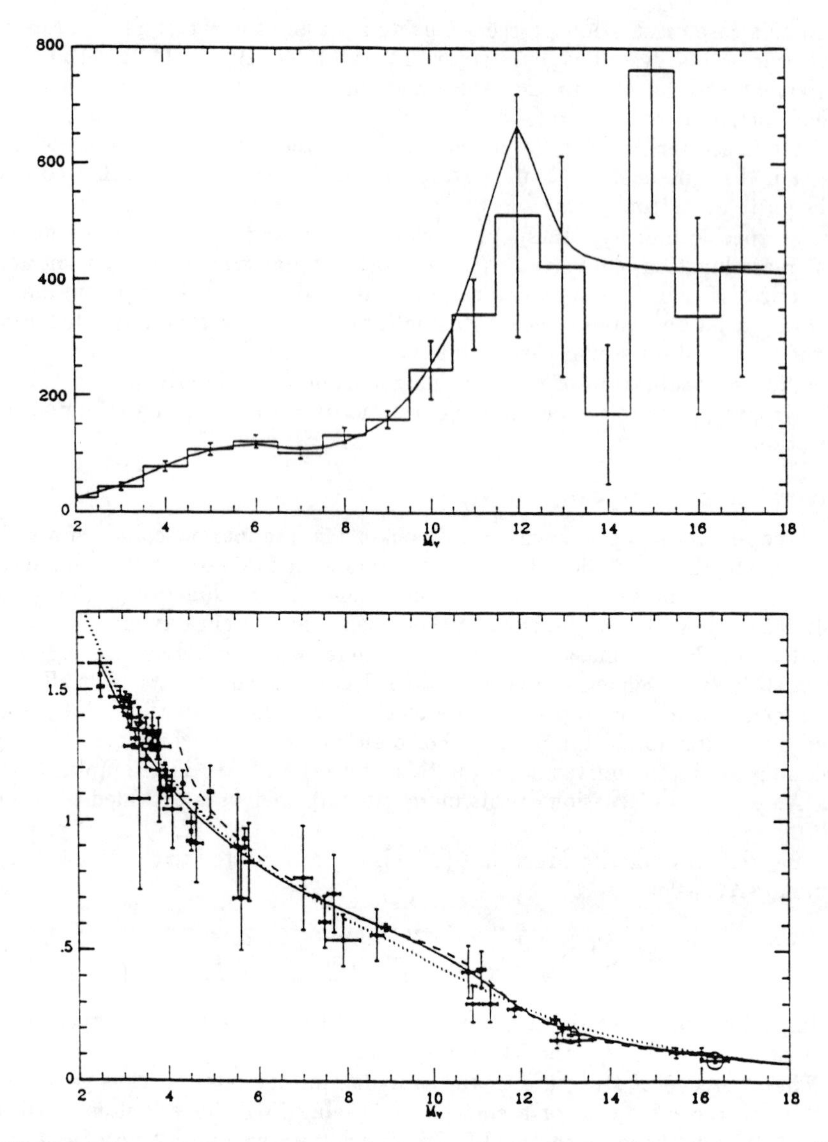

FIGURE Ia The upper panel shows the observed local luminosity function (Wielen, Jahreiss and Krüger 1983) extended beyond $M_v = 13$ as described by Kroupa, Tout and Gilmore (1990). The solid curve is the adopted single star luminosity function.

FIGURE Ib The lower panel shows the mass-luminosity relation, with data from Popper (1980) and Liebert and Probst (1987). The curves in the figure show various adopted relationships, with the solid curve that favoured in the analysis of Kroupa *et al.* .

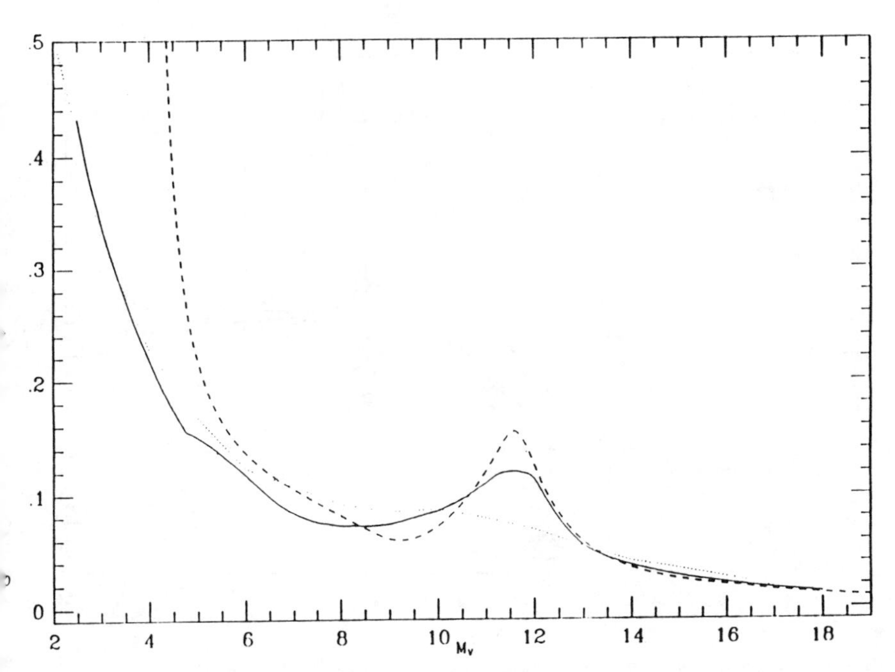

FIGURE II The absolute value of the slope, $d\mathcal{M}/dM_v$, of the mass-luminosity relation shown in Figure 1b. Note particularly how the shape of this function mimics that of the stellar luminosity function in Figure 1a.

function lacks structure. The structure in the mass-luminosity relation is understood in terms of changes in the opacity sources, and consequent changes in the equation of state, of stellar atmospheres. Further discussion is provided by Kroupa *et al.* (1990).

For modelling purposes, the relevant feature of the result shown in Figure 2 is that a *range* of mass functions and a *range* of mass-luminosity relations must be allowed for in the Monte Carlo process.

Monte Carlo Modelling - An Example

Given the preliminaries above we are now ready to begin. One wishes to include all the observational and physical effects discussed. In practice, one proceeds by Monte Carlo modelling. This rather formidable name means simply including random sampling into computer modelling. Statistical comparison of the range of results allowed by adjusting each parameter within its formal error range with the observational data then provides a statistically optimal value for each parameter, and an "error bar" about that value.

When a large number of parameters is present it is important to calibrate the method on a well studied case. In this example, the scatter in the H-R diagram for well studied trigonometric parallax stars is well reproduced. The many tests and checks are illustrated in the papers of Kroupa *et al.*, and need not be repeated here.

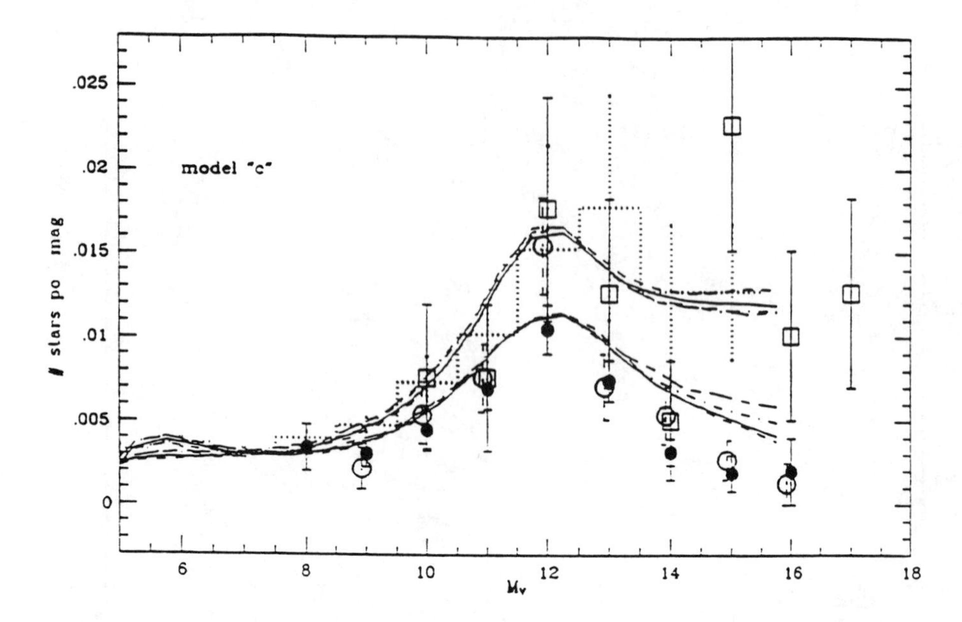

FIGURE III The single star luminosity function from Figure 1 (open squares) and two photographic system luminosity functions (Reid and Gilmore 1982, open circles; Stobie *et al.* 1989, solid circles). The solid curve is the single stellar IMF which is most consistent with all observational data, converted into the observational plane.

Rather, we summarise the results in Figures 3 and 4. Figure 3 shows the nearby star luminosity function, as in Figure 1, together with two determinations of the field star system luminosity function. Also shown is the resulting stellar mass function, converted into the observational plane, which is consistent with all available data on the absolute number of single stars near the Sun and their distribution with luminosity and colour, and with the apparent number of stellar systems to $\sim$ 100 pc from the Sun, and their distribution with colour and apparent magnitude. Figure 4 shows the stellar IMF which is consistent with all these data.

Is it worth it?

Following this extensive and detailed modelling, one wonders naturally if this is all necessary. At one level of course it is only when all known effects are included in models that one can be confident of the answers. Thus an explicit demonstration that the Malmquist corrections are well-understood and applied correctly is necessary so that more complex derived quantities can be derived. Examples include the stellar IMF discussed above, and the surface mass density of the Galactic disk derived by Kuijken and Gilmore (1989).

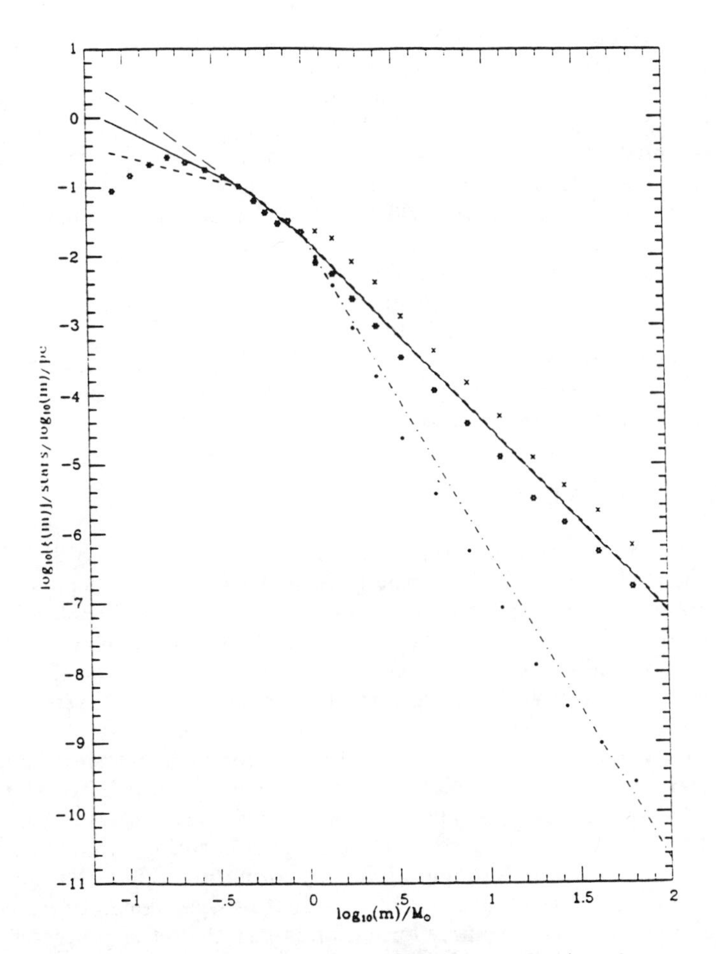

FIGURE IV The stellar initial mass function (solid line) and present day mass function (dash-dotted line) are most consistent with available data. The long dashed and short dashed lines at low masses illustrate the extreme range of likely results. The symbols near each line are the results derived by Scalo (1986). The only significant difference from those results is in the slope at very low masses.

As well as this, such modelling sometimes provides useful new predictions, or explanations of results which were not part of the original study. An example here is the apparent very steep decrease in the scale height of very late M stars reported by Hawkins (1988). He reports a decrease in disk apparent scale height from $\sim$ 200 pc for stars with $V-I \sim 2$ to only $\sim$ 75 pc for stars with $V-I \sim 3.5$.

This dramatic decrease is difficult to understand as a normal effect of the age-scale height relation for field stars, as it would require that almost all stars with $M_v \gtrsim +12$ have been formed in the last $\sim$ 1 Gyr. This result is rather an artefact of a bias towards equal masses in binary stars as one approaches the minimum mass for hydrogen burning.

If all binary stars are formed so that the two masses are chosen independently from the same IMF, then most secondaries will come from the most common stars, the late M dwarfs. If such a star has a massive primary companion it will have only a small effect on both the luminosity and colour of the system, and will be modelled as described earlier. A low mass primary however can have only a nearby equal mass *luminous* secondary. This directly gives a very large increase in the system luminosity, which will tend to a factor of two change at the lowest masses. [Lower mass companions of course have zero luminosity.]

This change provides a very considerable correction to the colour-magnitude relation, and leads to a considerable change in the apparent scale height of the distribution of stars with decreasing luminosity.

Interestingly, one could invert the process, and deduce the mass ratio distribution of the lowest mass stars from careful derivation of the scale height change. This remains to be attempted.

GALAXY FORMATION

Galaxy formation is simple in principle. One starts with a mass distribution and an associated velocity field at some time in the early Universe, and evolves it. The system will collapse round local over-dense regions, will cool, mix, form stars, and evolve. Rather like weather forecasting however, there are very many non-linear or poorly quantified processes happening, so that specific detailed calculations are unlikely to be either feasible or relevant to real systems in the foreseeable future.

A more tractable problem is to identify the dominant physical processes, the characteristic timescales, and the robust conclusions which are of general relevance. One of the most reliable methods to achieve this is to analyse the fossil record. In our case the fossils are those long-lived low mass stars which were formed during the early stages of Galaxy formation. Their relevant properties are their distribution functions in coordinate space, kinematic space, and chemical abundance space. Galaxy formation is just the set of processes which determines occupation numbers in those parts of phase space.

As a specific example, the spatial distribution of stars in the Galaxy is related to stellar kinematics by the gravitational potential. This allows two possible investigations. In cases where one can determine two of these distributions – kinematics and spatial distributions being the easiest – then the third can be measured. This is the basis of determinations of mass distributions in the Galactic disk and in the far halo.

Less rigorously, the *scale lengths* of a stellar system depend on how much the gas cloud from which the stars formed was able to cool and contract – that is, to *dissipate* its binding energy – before substantial star formation. The *shape* of a stellar system depends on a mixture of the large scale mass distribution, through the gravitational potential, and on the relative contributions of angular momentum and pressure to its support against the potential gradient. These in turn depend on the total collapse factor and the degree of torque-induced spin in the primordial system. Thus apparently simple properties of a galaxy - its size and shape – are themselves valuable fossil records of the conditions from which the galaxy formed.

A review of galaxy formation from an observation viewpoint is provided by Gilmore, Wyse, and Kuijken (1989), where more details may be found. We now present a summary of the essential features.

Timescales

There are three natural characteristic timescales which may be associated with the halo of a galaxy like the Milky Way. The first of these is the cooling time, which is the time for radiative processes to remove the internal energy of a cloud. Defining the cooling rate per unit volume to be $n^2\Lambda(T)$, where n is the particle number density, and where the form of $\Lambda(T)$ includes the contributions from free-free, bound-free, and bound-bound transitions, and thus is an implicit function of chemical abundance as well as being an explicit function of temperature T, gives

$$t_{cool} = \frac{3nkT}{n^2\Lambda(T)}. \tag{12}$$

The second natural timescale is the gravitational free fall collapse time of the system, which is the time it would take for the system to collapse upon itself if there were no pressure support. This depends only on the mean density of the system, and is

$$t_{ff} \sim 2\times 10^7 n^{-\frac{1}{2}}\ \mathrm{yr}. \tag{13}$$

The third timescale is the collision time between gas clouds in the halo, in the case where one imagines a halo to be made of N_{cl} independent clouds. This timescale is then a measure of the longevity of a cloudy halo against disruptive (and dissipative) cloud-cloud collisions. The cloud collision time is (York *et al.* 1986)

$$T_{coll} \sim \frac{1}{N_{cl}v\sigma} \sim 5\times 10^8 \left(\frac{R}{50\,\mathrm{kpc}}\right)^3 N_{cl}^{-1} \left(\frac{v}{100\,\mathrm{kms}^{-1}}\right)^{-1}\ \mathrm{yr}. \tag{14}$$

For the parameters of the halo of the Milky Way all three timescales are similar.

$$t_{cool} \sim t_{ff} \sim t_{coll} \lesssim 1\,\mathrm{Gyr}. \tag{15}$$

By comparison, the natural timescale for disk formation is $\sim$ 4 Gyr. This is deduced from the requirement that the disk mass must have fallen in from $\sim \lambda^{-1}\alpha_{disk}$ kpc, with α_{disk} the disk radial scale length ($\sim$ 4 kpc) and λ the disk angular momentum parameter ($\lambda = \ \sim .07$) together with simple calculations of the rate at which loosely bound material falls into a growing density perturbation in the early Universe. Thus theoretical prejudice leads to an expectation that Galactic halo formation was rapid, and could be complete by redshifts $\sim 2-3$, but disk formation should have continued until later times, $z \lesssim 1$.

Detailed Models

There are now many reasonably detailed modern calculations of the collapse of a protogalaxy. The essential features of these models are the treatment of the dark matter which dominates the gravitational potential, and hence the dynamics, and the treatment of the baryonic matter (hydrogen and helium) which forms the luminous galaxy. All models assume the dark matter interacts only through gravity, so its evolution can be followed in n-body dynamical experiments. Following the gas is more difficult, especially since there is no understanding at all of how and why star formation proceeds. The most detailed recent calculations use smoothed particle hydrodynamics methods, and are major research and computing projects. A good recent discussion is that by Katz and Gunn (1991), who show results which bear a close resemblance to real galaxies.

Some of the more important general features of the recent numerical models are interesting, in that they disagree with earlier theoretical expectations. One of the more general calculations which was widely used until recently related the cooling time for an efficiently virialised galaxy from equation 12 to the Hubble time to derive a characteristic galactic mass (Rees and Ostriker (1977)). While this produced results in agreement with observation, the numerical models show proto-galaxies built up in hierarchical galaxy formation models such as the modern cold dark matter (CDM) models, never approach their virial temperatures of order 10^6 K. Rather the gas is mostly near 10^4 K.

Another popular picture was that disks form from smoothly distributed gas, which had acquired some angular momentum from tidal torques, cooled and sank slowly to the centre of a stationary CDM halo. If the angular momentum distribution is conserved this process nicely produces thin exponential disks (e.g. Gunn 1987). The models however now show that disks form from the gas in the centres of CDM clumps, and that the disk and halo form at the same time. It may well be this vigorous shaking of the forming disk which created the thick disk.

There are now several calculations of galaxy formation with sufficient detail that one may begin serious observational tests. The most general is to establish the inter-relationships between the various definable stellar populations in the Galaxy, particularly to see if they are all aspects of a continuous sequence, or if there are definable discrete units in a galaxy which correspond to a specific event in its evolutionary history, for example a major merger.

Galactic Stellar Populations

The Galaxy is conveniently discussed in terms of discrete building blocks. It is very important to remember that the relationships between these units are our major topic of research in Galactic structure. The discreteness or otherwise of the stellar populations in the Galaxy is our primary evidence for the past history of Galactic evolution. Thus it is important not to *assume* a *physical* discreteness behind a division which is made mostly for convenience and historical reasons. With that caveat however, it is very often convenient to discuss the Galaxy in terms of components. These are typically the central bulge, the stellar halo (aka spheroid), the thick disk, the thin disk, the dark halo, and the interstellar medium (ISM).

These components can be represented in many parts of parameter space. Four such representations are shown in Figure 5.

The Outer Halo:

The spatial distribution of the outer galactic halo is poorly known. Available constraints come from modelling the kinematics of stars in the Solar Neighbourhood, and from in situ surveys. Each method has its strengths and weaknesses.

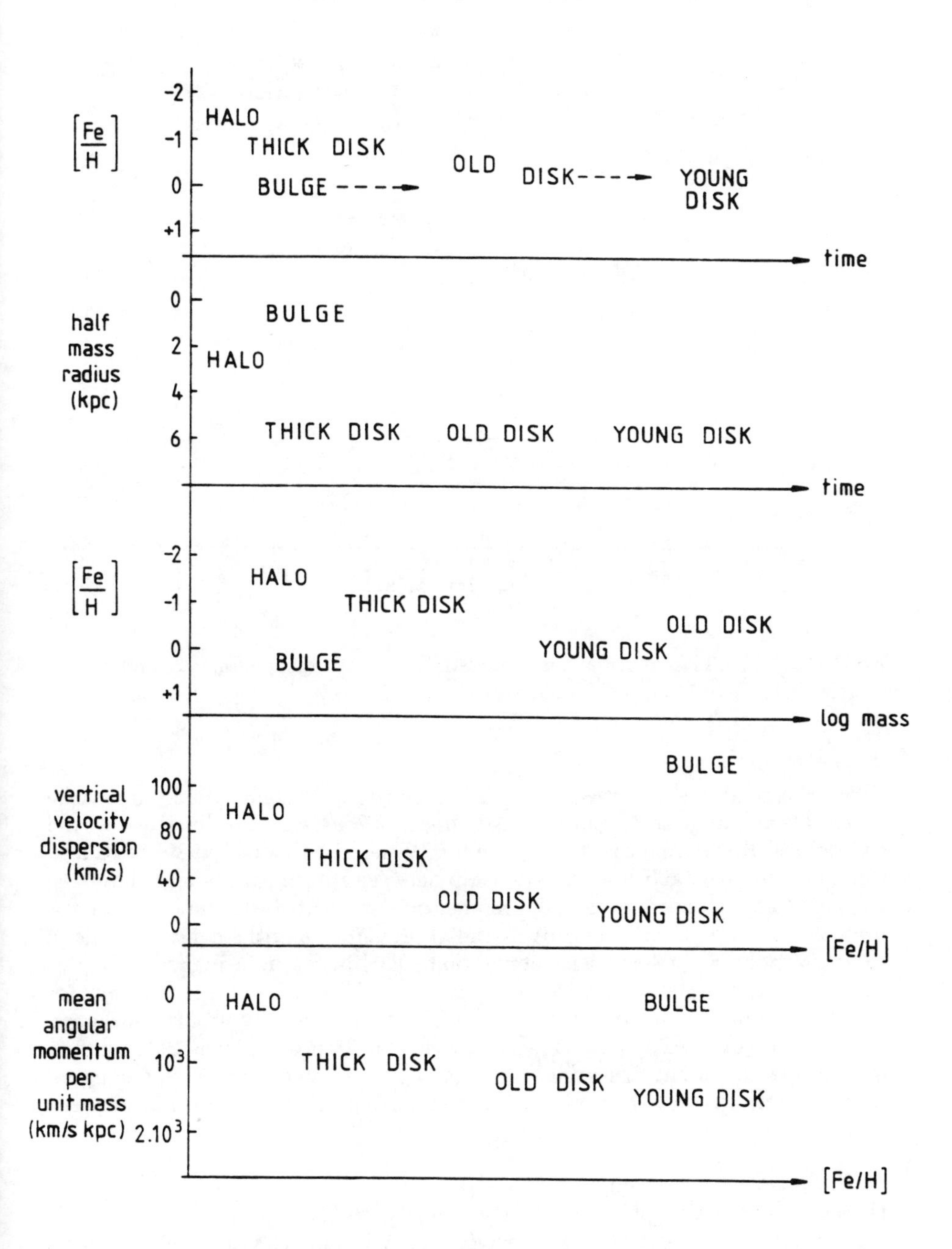

FIGURE V Different views of the Galaxy

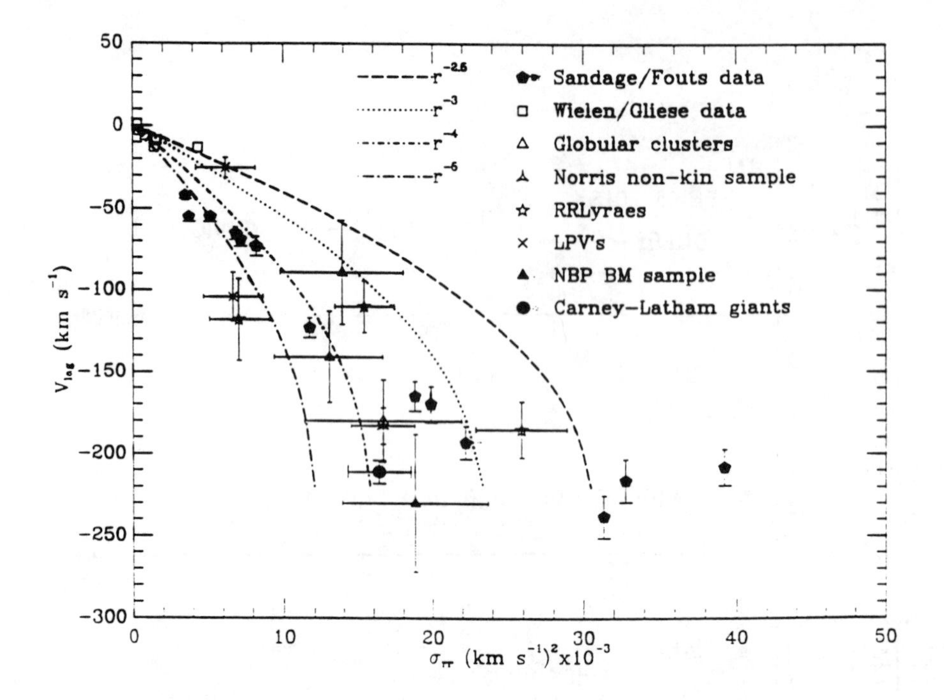

FIGURE VI The relation between radial velocity dispersion σ_{rr} and asymmetric drift V_{rot} of samples of old stars in the Galaxy.

A. Deduction from local kinematics.

There is a relationship between stellar kinematics at some location, large scale potentials and large scale spatial density distributions, which is described by the collisionless Boltzmann equation. Gilmore, Wyse, and Kuijken (1989) have used this equation to establish the relationship between rotational velocity about the Galactic Centre (effectively angular momentum), velocity dispersion radially out from the Galactic Centre (effectively radial pressure), and the density profile of the corresponding stellar tracer population. This is shown in Figure 6. In this figure, stellar samples binned in [Fe/H] are plotted, and show some tendency for the lowest abundance data to cross lines of constant density profile. Although this result is confused by possible selection effects (Ryan and Norris 1991), and so is of uncertain statistical significance, it is present in all current stellar samples. If real, it implies that the first stars formed during dissipational collapse of the Galactic halo.

B. In Situ Determinations

Direct determination of the outer halo density profile is complicated by small number statistics and distance uncertainties. A recent relevant result based on the spatial distribution of BHB stars has been derived by Arnold and Gilmore (1992). For a sample of BHB stars within $\sim$ 15 kpc of the Galactic Centre they derive a best fit power law index of the density law $\alpha = 3.4 \pm 0.2$, while for a more distant sample the result is $\alpha = 2.7 \pm 0.1$. [In both cases an axis ratio $q = c/a \lesssim 0.5$ is preferred.] These density profiles are in good agreement with the local kinematic results, in suggesting a shallow envelope in the outer Galaxy, and again marginally support the hypothesis that star formation began during dissipational collapse of the Galactic halo.

This result, albeit of low significance, means that the first star formation could have begun when the Galaxy first turned round from the general Hubble flow, and so is consistent with the first star formation occurring at very high redshifts. Even if star formation did occur during dissipation of the outer halo, this timescale deduction is of course not required. Nonetheless it does mean that in principle Galactic star formation can have occurred at the redshifts when the first QSOs are seen.

The Inner Halo

The very central regions of the halo are very highly dissipated. The bulge seen by IRAS (see for example Harmon and Gilmore 1988) has half-light radius of only a few hundred pc, smaller than the vertical extent of the disk. The relationship of these stars to the rest of the Galaxy remains unclear. They may be a central component of the old disk, a central component of the bulge, or a discrete structure (merger remnant?). The main *optical* bulge in the central few kpc of the Galaxy remains very poorly studied. Its properties are mainly deduced as yet by analogy with other galaxies (Wyse and Gilmore 1988). However, it does seem to be moderately metal enriched ($\gtrsim$ solar), moderately old, and to have a significant rotational contribution to its support against gravity. It is equally well explained as being the remnant of the first disk-like structure in the Galaxy, destroyed during later accretion, or as being evidence for a kinematic-chemical abundance gradient in the inner halo.

In either case the relevant result for the present discussion is that a metal enriched stellar population formed from moderately dissipated (the optical bulge) and from extremely highly dissipated (the IR bulge) gas at relatively early times.

Galactic Star Formation History

While absolute stellar age determinations remain difficult, differential measurements to determine the range of stellar ages in a narrow range of metallicities are more reliable. Such studies have been made for globular clusters (see Chaboyer *et al.* 1992 for a recent summary) and for field stars (Schuster and Nissen 1989). For globular clusters there is now a well established age range of ~ 3 Gyr, with the metal poor and inner clusters all being consistent with the same (old) age, and some of the outer clusters being younger. A similar result applies to local metal-poor field stars – an age range of a few Gyr is possible, though most of the subdwarfs are consistent with a narrow spread of (old) ages.

An adequate summary of the age and abundance information relevant for this discussion is that all halo star formation took place in $\lesssim 3$ Gyr some ~ 15 Gyr ago, with *most* in $\lesssim 1 - 2$ Gyr. The resulting abundance distribution is peaked at $\sim \frac{1}{30}$ the solar value.

Similar studies of disk stars however show a range of ages from nearly those of halo stars to stars forming now. The abundance of the ISM at the solar Galacto-centric radius was within a factor of 2 of solar some $\gtrsim 10$ Gyr ago, and has remained near solar since then.

The most direct limits on the star formation rate (SFR) in the early Galaxy come from chemical abundance considerations. Both the total abundance [Fe/H] and the individual element ratios are relevant.

Chemical Evolution of the Galactic halo

The halo stars at and beyond the solar galactocentric distance are adequately described as having a gaussian abundance distribution with mean [Fe/H] ~ -1.5 dex, and dispersion $\lesssim 0.5$ dex. Hartwick (1976) showed that this distribution, for an IMF like that of the disk (see below) requires that $\sim 90\%$ of the gas mass must have been lost. Wyse and Gilmore (1992) considered the angular momentum distribution of the various definable stellar populations in the Galaxy to show that the "lost" halo gas is most likely the precursor of the Galactic bulge.

This has an important consequence for chemical evolution models of the local disk which use the concept of the "Solar cylinder" – all stars in a column through the Sun are considered. A more physically well-founded sample should use a truncated cylinder, which should also curve as it rises away from the Plane, to allow for radial diffusion of stellar orbits. This is better described as a "Solar Sausage". The implication of this is that the chemical precursors of the Galactic disk population are still in the disk – the metal-weak thick disk discussed by the Mt Stromlo group may be of relevance here.

Element ratios and Star Formation Rates

Chemical element ratios are the most sensitive tracer of past star formation rates. This follows naturally from the different lifetimes of stars which produce different elements. The 'alpha' elements, especially oxygen, Ca, Mg, and Ti for present purposes, are created in short-lived ($\lesssim$ few$\times 10^8$ yrs) massive stars. Thus, the amount of oxygen created is a measure of the current star formation rate of massive stars. Other elements, including much of the iron, are created in longer-lived systems (Type I supernovae), which therefore reflect a time-averaged past SFR in their production rate. The element ratios in the ISM therefore reflect the ratio of the present SFR (oxygen) to the past averaged SFR (Fe). This ratio is preserved in newly forming stars, so that we can measure it now for the halo formation phase.

A detailed description of this type of model is presented by Wheeler, Sneden and Truran (1989) for our Galaxy, and explicitly in terms of limits on the past histories of star formation rates by Gilmore and Wyse (1991). The important observation is that the element ratio [O/Fe] is constant at a value ~ 0.55 throughout the whole halo abundance range. The most recent observational data extend this range to [Fe/H] $= -2.8$, beyond all but a few per cent of the mass of the halo (Gilmore, Edvardsson, Gustafsson and Nissen, 1992). Remarkably, the data show no detectable cosmic scatter in [O/Fe] over the whole range $-3 \lesssim$ [Fe/H]$\lesssim -1$, with a limit of $\lesssim 0.1$ dex resulting. Note that this range covers effectively the entire stellar mass in the halo.

The constancy and small scatter in this element ratio provide severe constraints on both the stellar initial mass function (IMF) and the SFR during halo formation. The constant value requires each forming star to see a well-mixed mass-averaged IMF of yield (see below). Thus, the mixing must have had time to be efficient, and the lowest mass stars contributing to the yield ($\sim 10\,M_\odot$)

must have had time to evolve. This sets an upper limit on the SFR which is hard to quantify, but based on guestimates of mixing times, local sound speeds, the extent to which SN occur in groups, and so on, the *maximum* SFR is $\lesssim 5$ times the mean SFR. The mean SFR for the halo is $\sim 10\,M_\odot$ yr^{-1}, to create $\sim 10^{10}\,M_\odot$ in stars in $\lesssim 10^9$ yr. Thus the proto-Galactic halo formation phase is unlikely to have been a starburst. Similar analyses for the Galactic disk, based on the systematic radial trends in [O/Fe] *vs* [Fe/H] and the very small scatter at any radius, similarly preclude any significant period in the disk with SFR more than a factor of a few higher than the present (radially-dependent) value (Edvardsson *et al.* in preparation).

Element Ratios and the Halo Stellar IMF
The numerical value of the [O/Fe] ratio for halo stars is determined by the stellar yields as a function of mass (and [Fe/H]) and by the relative number of stars as a function of mass, the IMF. Wyse and Gilmore (1992) have quantified the limits on the IMF provided by the abundance distribution of halo stars. Changes in the IMF are more reliably derived than is the slope, due to uncertainties in the theoretical yields. The Wyse and Gilmore calculation is summarised in Table 1. The result is that a systematic change in the value of [O/Fe] ~ 0.1, a realistic maximum allowed by observations, restricts the slope of the high mass IMF to have changed by less than ~ 0.4. Note that this limit *includes* any systematic metallicity-dependence of the yield. It is also interesting that the best match to the observations is with an IMF similar to that in the disk today. That is, remarkably, there is no evidence for any systematic change in the stellar initial mass function over the mass range $100\,M_\odot \gtrsim M \gtrsim M_\odot$ over 3 orders of magnitude in metallicity, $-3 \lesssim$ [Fe/H] $\lesssim 0$.

ASTROMETRY AND THE DARK HALO

The least understood major component of the Galaxy is its dark halo. Since this dominates the large scale gravitational potential, one can use dynamics to study the shape and extent of the dark matter. This may be done directly, by measuring the velocity dispersion and density profile in the outer halo (e.g. Arnold and Gilmore 1992). It is also possible to use high precision local kinematics.

It is quite likely that the dark halo is triaxial. Indeed, it is likely that the degree of triaxiality depends on the initial conditions for galaxy formation, so that a measure of triaxiality would be doubly informative. The shape of a CDM dark halo is discussed by, for example, Katz and Gunn (1991) and Dubinski (1992). The relevant point for now is that significant triaxiality is expected. This has an important consequence for the luminous galaxy which lives in the CDM potential, in that it can oscillate. Large-scale $m = 2$ mode oscillations (bar-like or triaxial) are allowed, and have recently been suggested to explain the kinematics of gas in the outer Galaxy (Blitz and Spergel, 1991).

Large scale oscillations have local consequences, which are potentially visible in stellar kinematics (Kuijken and Tremaine 1991). The most readily measurable effects arise in the radial velocity of the Local Standard of Rest (LSR), and in the orientation of the orbits of old stars. Old stars would, on average in a steady state Galaxy, have velocity vectors which point towards the centre of the potential, which is at the centre. In an oscillating system this is no longer true, and one might hope to see a derivation of the vertex of the average velocity ellipsoid away from the line to the Galactic centre. Old stars should also, in a steady state system, form a system which neither expands nor contracts, so that the radial velocity of the LSR should be zero. This need not be true in general.

The best way to check for either of these effects is to obtain astrometry of stars which are far from the Galactic plane. A crucial requirement is that the zero point of the measured kinematics be fixed with respect to objects which are unaffected by Galactic kinematics. Background galaxies and quasars are required. Given that, a radial motion of the LSR would be readily apparent in the mean motion of nearby stars, while the mean orientation of stellar orbits far from the plane is also measurable.

The most careful study of this type available is that by Soubiran (1991). She measured proper motions relative to background galaxies for a large sample of stars towards the north Galactic Pole. Her best determination for the mean motion of the LSR derived from nearby stars is $\bar{u} = 2 \pm 1$ km/s. Similarly, the mean difference between the vertex of the velocity ellipsoid of distant stars and the direction towards the Galactic centre is $6° \pm 6°$. For comparison, the Blitz and Spergel model predicts $\bar{u} = -14$ km/s and a vertex deviation of $-9°.3$

This illustration demonstrates both the information content of kinematics, and the way in which classical observational techniques may be utilised to test modern models.

CONCLUSION

The combination of modelling galaxies from first principles, to identify robust conclusions, and modelling observations in detail, to identify robust parameters, is the basis of astrophysics. We have much left to learn, but can take comfort in that we have at least an effective methodology available.

REFERENCES

Arnold, R. & Gilmore, G., 1992. *MNRAS* **257**, 225.

Blitz, L. & Spergel, D., 1991. *ApJ* **370**, 205

Chaboyer, B., Sarajedini, A. & Demarque, P., 1992. *ApJ* **394**, 515.

Dubinski, J., 1992. *ApJ* (in press).

Gilmore, G., Wyse, R.F.G., & Kuijken, K., 1989. *ARA&A* **27**, 55

Gilmore, G., & Wyse, R.F.G., 1991. *ApJ* **367**, L55

Gilmore, G., Gustafsson, B., Edvardsson, B., & Nissen, P.E. 1992, *Nature* **357**, 379.

Gunn, J.E., 1987, in "The Galaxy", ed. G. Gilmore & R. Carswell (Dordrecht: Reidel) p413.

Harmon, R.T. & Gilmore, G., 1988. *MNRAS* **235**, 1025.

Hawkins, M.R.S. 1988. *MNRAS* **234**, 533.

Hartwick, F.D.A., 1976. *ApJ* **209**, 418.

Katz, N. & Gunn, J.E. 1991. *ApJ* **377**, 365.

Kroupa, P., Tout, C.A. & Gilmore, G. 1990. *MNRAS* **244**, 76.

Kroupa, P., Tout, C.A. & Gilmore, G. 1991. *MNRAS* **251**, 293.

Kroupa, P., Tout, C.A. & Gilmore, G. 1992. *MNRAS* (in press).

Kuijken, K. & Tremaine, S. 1991 in *"Dynamics of Disk Galaxies"* ed. B. Sundelius (Goteborg Univ) p71.

Liebert, J. & Probst, R., 1987. *ARA&A* **25**, 473.

Lutz, T.E. & Kelker, D.H. 1973. *PASP* **85**, 573.

Popper, D.M. 1980. *ARA&A* **18**, 115.

Rees, M.J. & Ostriker, J.P. 1977. *MNRAS* **179**, 541.

Reid, I.N. & Gilmore, G. 1982. *MNRAS* **201**, 73.

Ryan, S. & Norris, J.E. 1991. *AJ* **101**, 1835.

Scalo, J.M. 1986. *Fund. Cosmic Physics*, **11** 1.

Schuster, W. & Nissen, P.E. 1989. *A&A* **222**, 69.

Soubiran, C. 1991. *Thèse de Doctorat*, Paris.

Stobie, R.S., Ishida, K. & Peacock, J. 1989. *MNRAS* **238**, 709.

Wheeler, J., Sneden, C., Truran, J.W. Jr. 1989. *ARA&A* **27**, 279.

Wielen, R., Jahreiss, H. & Krüger, R. 1983 in *"Nearby Stars and the Stellar Luminosity Function"* ed. A. Davis Philip & A. Upgren (Davis Press, NY) p163.

Wyse, R.F.G., & Gilmore, G., 1988. *AJ* **95**, 1404.

Wyse, R.F.G., & Gilmore, G., 1992. *AJ* **104**, 144.

York, D.G., Dopita, M., Green, R., & Bechtold, J. 1986. *ApJ* **311**, 610

The Minnesota Lectures on Clusters of Galaxies and Large-Scale Structure
ASP Conference Series, Vol. 39, 1993
Roberta M Humphreys (ed.)

THE CENTRAL TEN PARSECS OF OUR GALAXY

John H. Lacy
Department of Astronomy
University of Texas
Austin, TX 78712-1083

I. INTRODUCTION

Galactic nuclei are well known to be the locations of some of the most interesting and exotic objects in the Universe. Unfortunately, at the distances of even the closest external galaxies, the region containing these objects is almost unresolvable. At the distance of M31, a parsec subtends only a few tenths of an arcsecond. Only in our own Galactic center, at a distance of ~ 8 kpc, where a parsec subtends ~ 25", can we hope to study the contents of a galactic nucleus in detail.

The drawback to studying the nucleus of our own Galaxy is that we must observe it through the dust in the intevening disk, which causes ~ 30 mag of visual extinction. As a result, the Galactic center has been almost exclusively the subject of infrared, radio, and occasionally γ-ray observations.

The Galactic center has been the subject of numerous reviews (*e.g.*, Oort 1977; Brown and Liszt 1984; Genzel and Townes 1987) and symposia (Riegler and Blandford 1982; Backer 1987; Morris 1989). In this review I do not attempt to duplicate the very thorough treatment given the subject in the past. Instead I restrict myself to a discussion of the work done since the Genzel and Townes review, and for the most part I restrict myself to the inner 10 pc (4') region. As this is the region which is unresolvable in external galaxies, it is perhaps the most important region to study in our own Galactic nucleus. Even with these restrictions, I cannot cover all work that has been done. I try to summarize the work I found most interesting, with a particular orientation toward work related to the mass distribution in the inner 10 pc and the question of a central black hole.

This review consists of two parts. The first is largely morphological, discussing the contents of the Galactic nucleus and how they are distributed. I start with the stars, then discuss molecular gas, atomic gas, and dust. The second part concentrates on the kinematics and inferred mass distribution. Again, I first discuss the stars and then the various gas components. Finally, I mention some of the uncertainties involved in determining the mass distribution, as well as non-kinematic evidence for and against the presence of a massive black hole.

II. MORPHOLOGY

Stars

Although I am concentrating on the central 10 pc, it is useful to tie the innermost region into the rest of the nucleus by discussing first the stellar distribution in the inner few hundred pc. Several pictures have been published of the near-infrared starlight from the nuclear region. Glass *et al.* (1987) made J, H, and K-band images of the inner 1 x 2° (150 x 300 pc) by scanning a single pixel detector. Gatley *et al.* (1989) imaged the inner 30 x 40' (75 x 100 pc) at K band with a PtSi array. Both data sets show the central star cluster, elongated along the Galactic plane, and the effects of dust, which still causes patchy extinction at 2.2 μm.

There seems to be a lack of published images with field sizes between 1' and 1°. Several images exist of regions ~ 1' in size. DePoy and Sharp (1991) imaged the inner 1 pc at J, H, K, and L. Their K-band image is shown in Figure 1. Outside of ~ 0.1 pc (2.4"), the images show a cluster of late-type stars. The IRS 16 cluster is seen in the inner 0.1 pc. It consists of hotter stars, which don't show 2.3 μm CO bandhead absorption. Eckart *et al.* (1992) used shift-and-add techniques to obtain a 0.25" resolution image at the IRS 16 cluster. At this resolution the cluster splits up into about two dozen sources. Notably, one of the fainter sources is coincident, within 0.1", with Sgr A*, the radio source often considered to be the actual center.

Allen, Hyland, and Hillier (1990) combined imaging and spectroscopy in the K band and in Br α to study the central region. They found a number of point sources of broad He I (2.06 μm) emission, which they identified as hot mass-loss stars (WN9/Ofpe).

Krabbe *et al.* (1991) imaged the inner 45" (2 pc) diameter in the 2 μm region through a Fabry-Pérot interferometer to study the distribution of the He I (2.06 μm) radiation. They confirmed the conclusion that IRS 16 consists of mass-loss stars, explaining the broad He I emission, previously suggested to be evidence of low-level Seyfert activity. Their continuum and line images are shown in Figure 2.

Stellar Distribution

Rather little has changed in our knowledge of the distribution of starlight in the Galactic nucleus since Becklin and Neugebauer (1968) found the 2.2 μm surface brightness to vary approximately as $r^{-0.8}$. This requires a luminosity density varying as $r^{-1.8}$, or falling slightly more gradually than expected in an isothermal star cluster with constant M/L. The only major question has been about the core radius of the light distribution. Allen *et al.* (1983) argued that the $r^{-0.8}$ distribution extends in to 1" (0.04 pc) from the center. However, Rieke and Lebofsky (1987) found that the faint light between the bright stars levels off at r ~ 0.8 pc. The finding by Allen, Hyland, and Hillier that the inner region is populated by unusual hot stars suggests that

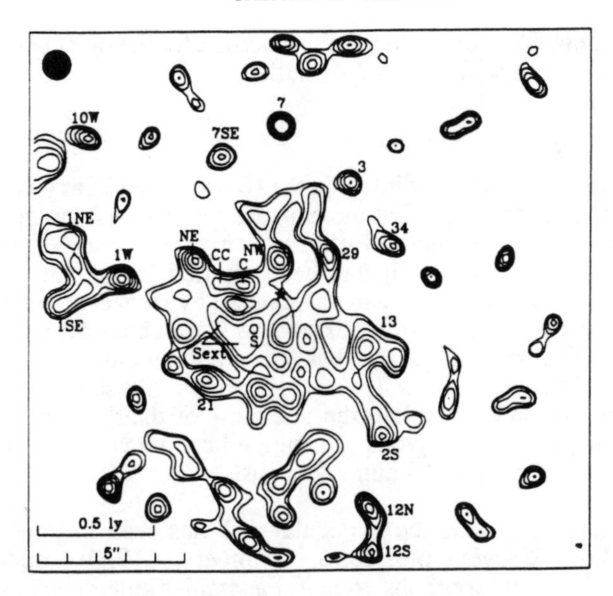

Figure 1. Contour map of K-band image of the central 1 x 1 pc of the Galaxy. Taken from DePoy and Sharp (1991).

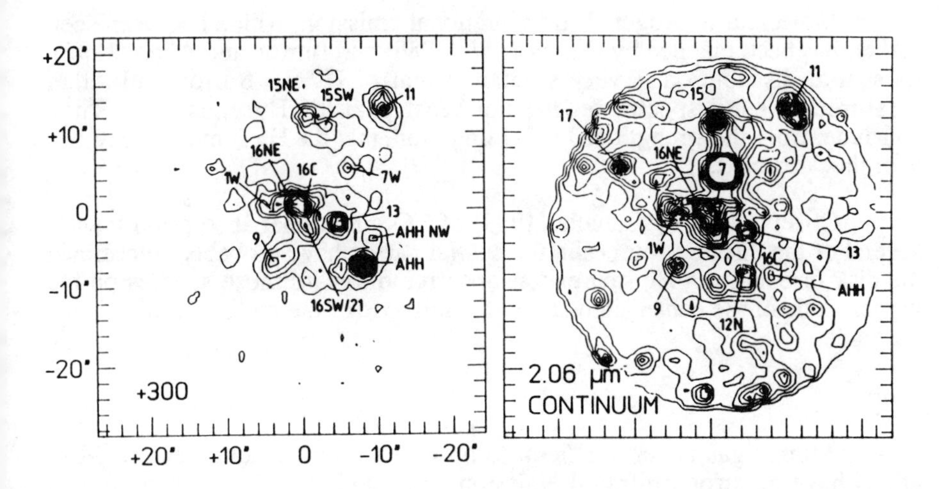

Figure 2. Contour maps of 2.06 µm continuum and He I line emission from the central 2 x 2 pc of the Galaxy. The line map covers a 300 km s^{-1} velocity range centered at +300 km s^{-1}. Taken from Krabbe *et al.* (1991).

the Rieke and Lebofsky core radius may be correct for the older stars which are more representative of stellar mass distribution.

Molecules

The majority of the gas in the inner 10 pc of the Galaxy is in the form of a molecular disk or ring. The molecular disk was first suggested by the double-lobed structure of the 100 μm dust emission observed by Becklin, Gatley, and Werner (1982). It has been studied in the lines of various molecules and is discussed in considerable detail by Genzel and Townes. The molecular gas forms a flaring disk extending from about 10 pc radius in to about 1.5 pc. The motions in the disk are predominantly circular with an approximately flat rotation velocity of ~ 110 km s^{-1}. Substantial turbulence is seen, however, which can explain the thickness of the disk, and anomalous velocities are seen in some regions. I mention here a few recent observations that have added to our understanding of the disk.

In the last few years the molecular disk has been mapped in excited rotation state lines of several molecules. Sutton *et al.* (1990) observed CO J = 3-2 emission. Their observations argue for a simpler kinematic description, a flat 110 km s^{-1} rotation, than inferred from some earlier measurements. Jackson *et al.* (1992) observed HCN J = 3-2 emission. This line requires relatively high densities to be excited, and peaks near the inner edge of the disk, at r ~ 2 pc. In contrast to earlier HCN J = 1-0 observations, they observed a complete ring of gas centered quite close to Sgr A*. They suggested that the disk consists of several somewhat distinct streamers.

Molecular hydrogen 2 μm vibrational emission, which had been seen earlier, has been mapped by DePoy *et al.* (1989) and Burton and Allen (1992). Its spatial distribution is very similar to that of HCN. Burton and Allen measured K band spectra on the northeast peak of H_2 emission. They conclude, from the strength of J = 2-1 emission, that the H_2 is more likely UV than shock excited.

Observations by Geballe (1989) of CO vibrational absorption toward several of the infrared sources in the central parsec have probably determined the disk orientation. He saw excess absorption toward those sources on the western side of the center, indicating that this side of the molecular disk is the near side.

Atoms

Atomic gas is very difficult to observe in the Galactic center. Most atoms have no strong infrared emission lines, and H I 21 cm absorption is quite weak from gas at the temperatures expected. Observations of [O I] 63 μm emission show that atomic gas is present, however. Jackson *et al.* (1992) mapped the [O I] emission with 20" resolution. They found that the atomic gas fills in the hole in the molecular disk, and has a distribution and kinematics like the ionized gas, described below. The required mass of atomic gas is ~ 300 $M_{\odot}$ within r = 1.5 pc, a comparable mean surface density to that of the

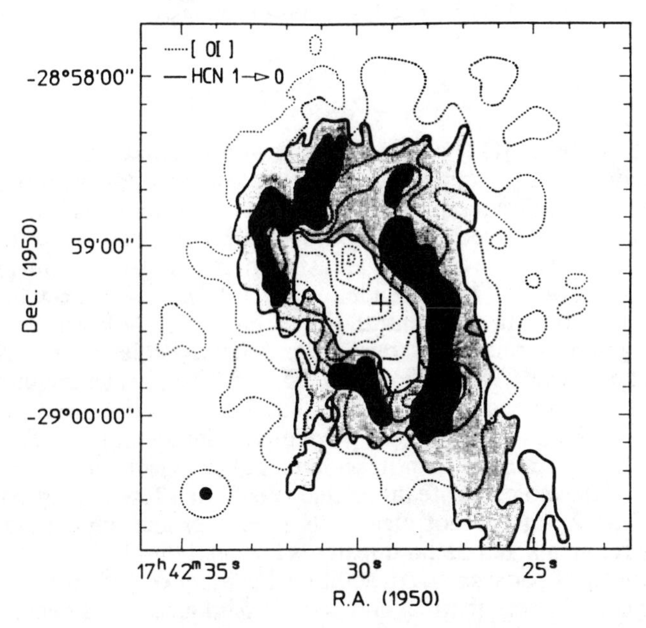

Figure 3. Contour map of [O I] (63 μ) emission superimposed on HCN J = 1-0 map at the central 8 pc of the Galaxy. Taken from Jackson *et al.* (1992).

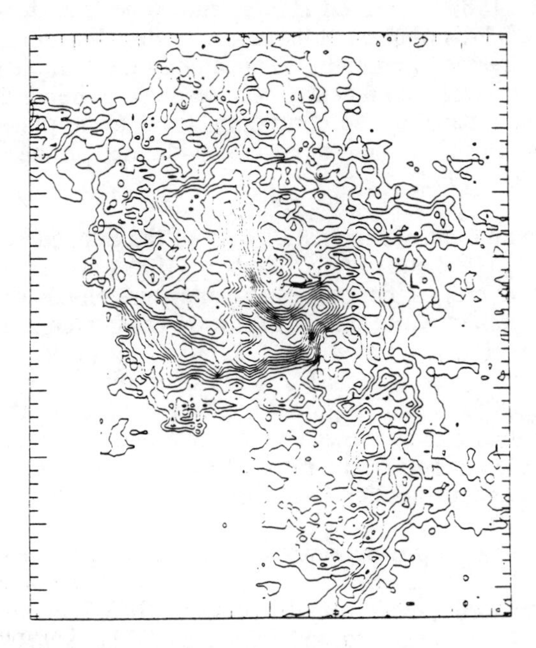

Figure 4. Contour map of [Ne II] (12.8 μm) emission from the central 3 x 4 pc of the Galaxy. Taken from Lacy *et al.* (1992).

molecular gas. The [O I] map, superimposed on HCN 1-0, is shown in Figure 3.

Ions

Although the [O I] observations have now shown that the majority of the gas inside the 1.5 pc inner edge of the molecular disk is neutral atomic gas, the most easily and frequently observed gas in this region is ionic. The VLA has been repeatedly used for progressively better imaging of the Galactic center. Two very nice images are shown in the IAU #136 proceedings (Killeen and Lo 1989; Yusef-Zadeh *et al.* 1989). They show detailed sub-arcsecond structure in the central H II region, Sgr A West, as well as the neighboring non-thermal shell source, Sgr A East. Pedlar *et al.* (1989) imaged Sgr A with constant 6" resolution at 6, 20, and 90 cm wavelengths to separate the various components. They found that Sgr A West is seen in absorption against Sgr A East at the longest wavelength, indicating that East is behind West. How far behind is not known, but it has been argued from its association with molecular features that it is with a few tens of parsecs. The nature of Sgr A East is not clear. Its spectrum and appearance resemble supernova remnants, but its luminosity is unusually high. It may be related to nuclear activity of some sort. An additional component, referred to as the halo, appears to contain both thermal and non-infrared gas, and is centered near Sgr A West.

Radio recombination lines have been mapped by several groups. Schwarz *et al.* (1989) observed H76α, and Roberts *et al.* (1991) observed H92α. The spatial distribution of the recombination line emission is much like that of the free-free continuum, but is apparently weak in the region just south of Sgr A*. This puzzle has recently been resolved as being due to inadequate spectral coverage causing the continuum level to be placed too high, the correlation with position being due to the wide lines in the innermost region (Roberts and Goss 1992).

A large amount of kinematic study of the ionized gas has been done with the [Ne II] (12.8 μm) line. Serabyn (1989) argued from the [Ne II] velocities that several of the ionized streamers are clouds of gas which have been tidally stretched while falling in toward the center, and ionized by UV radiation there. Lacy, Serabyn, and Achtermann (1991) reconsidered the ionized gas morphology, using a 1", 30 km s^{-1} resolution complete map of the inner 75 x 90" (3 x 4 pc) (Figure 4). They concluded that the two most prominent features can be viewed as a single one-armed spiral in the same plane as the molecular disk. The infrared hydrogen recombination lines have essentially the same distribution (and kinematics) as [Ne II].

A somewhat surprising ionized gas source was noticed by Yusef-Zadeh and Morris (1991) and Serabyn *et al.* (1991). They found a tail of ionized gas extending to the north (away from the center) of IRS 7, the most luminous red supergiant in the nuclear region (Figure 5). The Serabyn *et al.* [Ne II] observations show the gas velocity on IRS 7 to match the stellar CO velocity, and the gas velocity in the tail to shown gradual acceleration. Yusef-Zadeh and Melia (1992) made higher resolution VLA observations of IRS 7, and found a

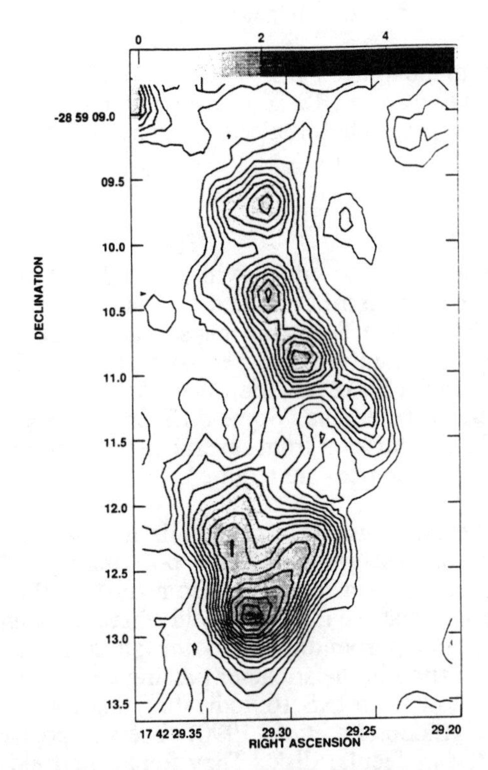

Figure 5. Contour map of 15 GHz continuum emission from the ionized tail of IRS 7. Taken from Yusef-Zadeh and Melia (1992).

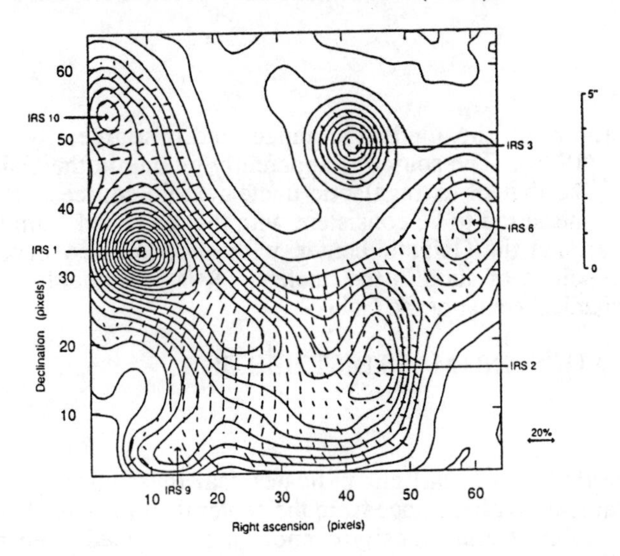

Figure 6. Polarization vectors of the 12.4 μm dust continuum emission superimposed on continuum contour map. Taken from Aitken *et al.* (1991).

bow-shock morphology to the ionized gas. The most likely explanation of the observations is that the wind from IRS 16 is sweeping back the slow neutral wind from IRS 7, and the gas is photoionized, perhaps by UV radiation from IRS 16.

Dust

Infrared emission from dust mixed with ionized and neutral gas can also be used as a tracer of the distribution and excitation of the gas. Smith, Aitken, and Roche (1990) made 8-13 μm maps and spectra of the central parsec. The distribution of mid-infrared dust emission is much like that of the ionized gas, since it is trapped Ly α radiation that heats dust to temperatures sufficient to emit at these wavelengths. Some differences are seen, however, notably excess mid-infrared emission from several compact sources. Smith *et al.* conclude that local heating sources are required.

A particularly interesting use of infrared emission from dust is to study the magnetic alignment of dust grains through polarimetry. Aitken *et al.* (1991) observed polarized 12 μm emission from dust in the ionized gas (Figure 6). They concluded that the magnetic fields there are aligned with the gas streamers (polarization perpendicular to the streamers). They also concluded that the polarization in the streamer passing closest to the center is not perturbed by the outflow from IRS 16, indicating that the flow is in front of or behind the center. Hildebrand *et al.* (1990) observed polarized 100 μm emission from dust in the molecular disk. They found the field to lie in the plane of the disk, and suggested a model in which differential rotation of partially ionized gas in the disk could drag initially poloidal field lines into the disk.

*Sgr A**

Observations of the nuclear compact radio source, Sgr A*, are reviewed by Lo (1989). The source is apparently unique in the Galaxy, but has a spectrum like that of extragalactic nuclear radio sources. Its proper motion has been measured to be consistent with that expected from the sun's orbital motion around the Galactic center, with uncertainties several times smaller than velocities of stars in the nucleus. It is presumably a massive object at the dynamical center of the Galaxy.

III. KINEMATICS AND MASS DISTRIBUTION

Stars

The distribution of starlight indicates that the stellar density falls somewhat less quickly with distance from the center than in an isothermal star cluster. The absolute stellar density cannot be determined from the light distribution without accurate knowledge of the stellar population. In fact, variations in the stellar population with distance could even invalidate conclusions about the relative density. Worse yet, some part of the mass may be contained in a dark component with a distribution different from that of the stars. A better determination of the mass distribution can be obtained from the kinematics of the star which should respond to all forms of mass.

The kinematics of the stars in the inner bulge can be studied with OH/IR stars. Lindqvist *et al.* (1989) measured velocities of 125 OH/IR star within 120 pc of the Galactic center. The velocities are characterized by a dispersion of $\sigma \approx 90$ km s^{-1} plus a gradient of $dv/dl \approx 120$ km s^{-1} kpc^{-1}. The determination of a mass distribution from the Doppler shifts of a group of stars is not simple, since only one velocity component is measured and the orbits of the star are not known. Lindqvist *et al.* used a method proposed by Bahcall and Tremaine (1981), which assumes that the stellar velocity dispersion is isotropic, to estimate the mass distribution. They concluded that a distribution of $M(r) = 4 \times 10^6 \, r_{pc}^{1.4} \, M_\odot$ for 10 pc < r < 100 pc is consistent with their data.

The kinematics of the stars in the inner few parsecs has been investigated, using the 2.3 μm CO bandhead in red giants, by Rieke and Rieke (1988) and Sellgren *et al.* (1990; see also Sellgren 1989). A much steeper velocity gradient is seen in their region than farther out, $dv/dl \approx 30$ km s^{-1} pc^{-1}, indicating substantial rotation of the nuclear star cluster. Sellgren *et al.* found evidence for an increase in the dispersion from $\sigma \approx 60$ km s^{-1} at $r \gtrsim 2$ pc to $\sigma \approx 125$ km s^{-1} at $r \lesssim 0.6$ pc (Figure 7). The dispersion levels off inside 0.6 pc, but the strength of the CO bandhead does too, indicating a lack of red giants in this region. They suggested that either red giants are destroyed inside 0.6 pc or the CO in their atmospheres is destroyed, and that the CO bandhead seen in projection inside this radius is actually in stars at greater true distances from the center. The more recent observations of a cluster of blue mass-loss stars in the inner region may supply a natural explanation if the population containing the red giants has a core radius ~ 0.6 pc. So long as the red giants which Sellgren *et al.* and Rieke and Rieke use as kinematic tracers of the mass distribution have core radius ~ 0.6 pc, Sellgren *et al.* concluded that additional dark mass of ~ $3 \times 10^6 \, M_\odot$ must be present in the core. The required distributed mass is $M(r) \approx 3 \times 10^6 \, r_{pc} \, M_\odot$.

Molecules

The rotation pattern of molecular clouds can be used to probe the mass distribution between ~ 1.5 pc and 1 kpc (and beyond). Binney *et al.* (1991) modeled observations of CO, CS, and H I from the inner 10° of the Galactic disk. They concluded that they could explain the observations if the gas orbits in the potential of a barred disk with the bar oriented 16° from our line-of-sight. They required a mass distribution of $\rho(r) \propto r^{-1.75}$ ($M(r) \propto r^{1.25}$) inside of $r \lesssim 1.3$ kpc. Notably, this radial dependence is very close to that of the 2 μm starlight ($L(r) \propto r^{1.2}$) and that inferred from the OH/IR star kinematics ($M(r) \propto r^{1.4}$).

At smaller radii ($r \lesssim 10$ pc), the molecular disk provides information about the mass distribution. It has a rotational velocity of $v \approx 110$ km s^{-1}, implying $M(r) \approx 2.8 \times 10^6 \, r_{pc} \, M_\odot$ between $r \approx 1.5$ and 8 pc. Like the mass inferred from the Galactic disk gas, this mass distribution is entirely consistent with that derived from stellar kinematics.

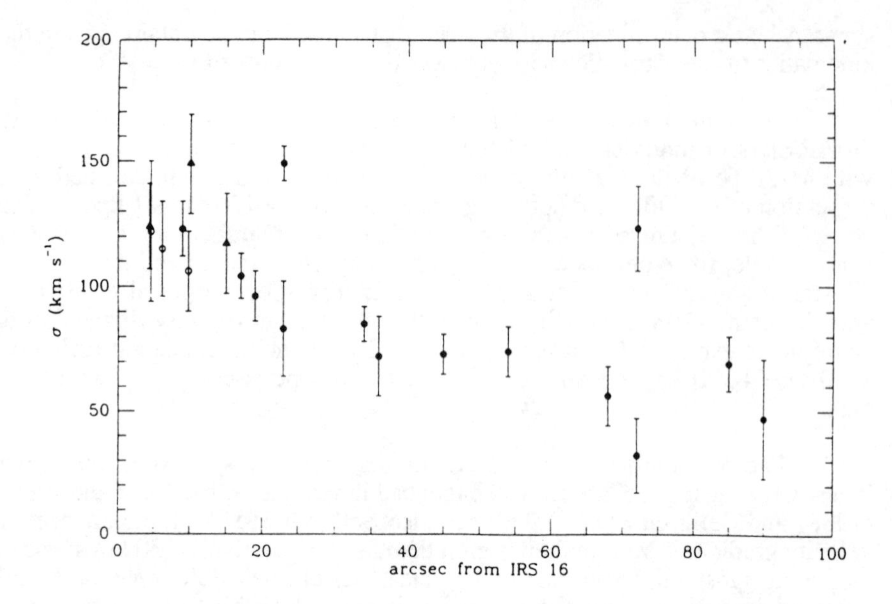

Figure 7. Stellar velocity disperson versus Galactocentric radius for 4" < r < 90". Taken from Sellgren *et al.* (1990).

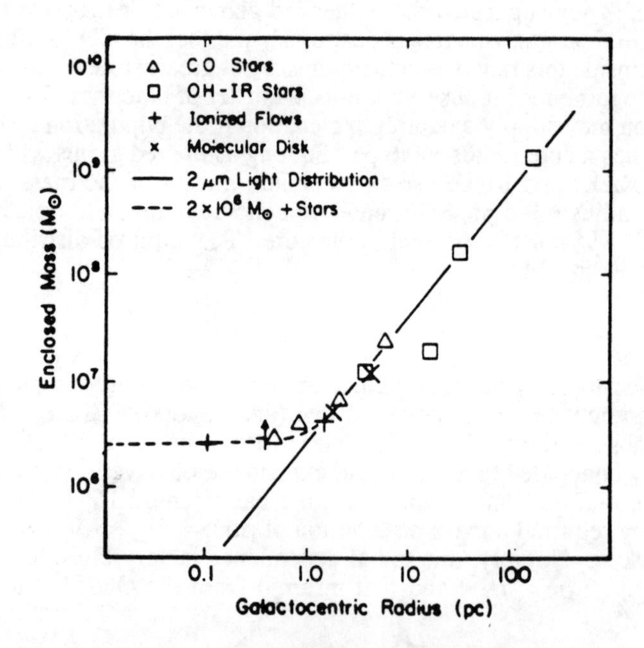

Figure 8. Enclosed mass versus Galactocentric distance. Points are taken from various observations of stars and gas.

Ionized Gas

The first kinematic determination of the mass distribution in the inner few parsecs was through infrared observations of [Ne II] (12.8 μm) emission from ionized gas. Serabyn (1989) summarized the conclusions from observations up to that time. He made models for gas flows along a ~ 1.5 pc radius arc (the "western arc") interpreted as the ionized inner edge of the molecular disk and a streamer (the "northern arm") extending in from the western arc to a point ~ 0.4 pc behind the center. He concluded that 4 x 10^6 $M_\odot$ is required inside of 1.5 pc and 3 x 10^6 $M_\odot$ is required inside of 0.4 pc radius.

More recent observations by Lacy, Achtermann, and Serabyn (1991) suggest that the western arc and the northern arm are a part of a single spiral extending from $r \approx 2$ pc in to $r \approx 0.1$ pc. They concluded that the gas does not flow along the spiral, but moves on nearly circular orbits. The rotation curve matches that of the molecular disk at $r \approx 1.5$ pc, but increases in a Keplerian fashion inside this radius, rising to $\approx$ 400 km s^{-1} at $r \approx 0.1$ pc. The mass distribution which best fits the ionized gas kinematics includes a 2.5 x 10^6 $M_\odot$ central mass and 1 x 10^6 r_{pc} $M_\odot$ distributed mass, but the observations are not very sensitive to the distributed mass, as the central mass dominates over most of the region observed. If a distributed mass of 2.5 x 10^6 r_{pc} $M_\odot$, consistent with other observations, is assumed, a central mass of 2.0 x 10^6 $M_\odot$ is required.

The ionized gas has also been observed in infrared and radio hydrogen recombination lines. The line shapes of the recombination lines are consistent with those of the [Ne II] lines, although the recombination-line observers disagree with some of the interpretations of Lacy, Achtermann, and Serabyn (Roberts and Goss 1992).

Mass Distribution Model

Although there are possible pitfalls in using any one of the kinematic tracers of the mass distribution, all of the observations now apper to be consistent with a single mass distribution, suggesting we are at least close to the resolution of this question. The distributed mass follows $M(r) \approx 3 \times 10^6$ $r_{pc}^{1.2}$ $M_\odot$ for 1 pc $\lesssim r \lesssim$ 1 kpc. As the starlight follows nearly the same law it seems likely that this distributed mass consists of stars, and that the mass-to-luminosity ratio is relatively constant throughout this region. There is evidence of a core radius ~ 0.6 pc for the distributed mass. Inside of that radius, the starlight probably does not trace the mass. Finally, a central mass, presumably a black hole, of 2-3 x 10^6 $M_\odot$ is indicated both by the gas kinematics and the stellar kinematics. The various kinematic data on the mass distribution are summarized in Figure 8.

IV. UNCERTAINTIES AND DISCUSSION

Some of the uncertainties in the kinematic determinations of the mass distribution have already been mentioned. In particular, the primary difficulty in interpreting stellar velocities is the possibility of anisotropic dispersion. At present, there are no observations which measure the velocity anisotropy, although high resolution HST infrared observations should be able to measure proper motions of stars at the Galactic center, providing three-dimensional velocities for the red giants, and measuring motions of the inner cluster of blue stars. In fact, ground-based observations should be able to detect proper motions of the inner cluster with a ~ 10 year baseline if they move at the expected 400 km s^{-1} velocities.

Non-gravitational forces are the primary traps in interpreting gas kinematics. The possibility of ejection, which could result in velocities unrelated to the gravitational potential, is ruled out by observations of [Ne II] Doppler shifts characteristic of nearly circular orbits. But magnetic fields of gaseous drag could significantly affect the gas motions.

Magnetic Fields

The importance of magnetic fields can be estimated by comparing the magnetic energy density (or pressure) to the kinetic energy density of the gas. Both are uncertain, but estimates can be made.

Aitken *et al.* (1991) suggested that since the classical Davis-Greenstein grain-alignment arguments tend to overestimate magnetic fields, their polarization observations could only supply a lower limit to the field strength based on the uniformity of the polarization. They concluded that the field strength is at least a few milliGauss. Based on similar arguments, Morris (1990) argued that B ~ 1 mG throughout the inner ~ 70 pc, and that on that large scale, the field is predominantly poloidal.

More direct measurements of the magnetic field should be possible using the Zeeman affect. Roberts *et al.* (1991) searched for Zeeman splitting of the H92α line, and put an upper limit to the line-of-sight field component in the northern arm of 15 mG. Unfortunately, they were most sensitive to Zeeman splitting in regions of small Doppler shift, where the gas motions, and so probably the field, are across our line of sight. Killeen *et al.* (1990) and Schwarz and Lasenby (1990) observed Zeeman splitting of molecular lines in the molecular disk. The concluded that $\langle B_{||} \rangle \sim 1$ mG.

It seems likely from the various observations that the field in the inner few parsecs is somewhere in the range of 1-10 mG. This gives an energy density of $B^2/8\pi = 4 \times 10^{-8} - 4 \times 10^{-6}$ erg cm^{-3}. The kinetic energy density depends on the gas density. The density in clumps in the molecular disk is ~ 10^5 cm^{-3} (Townes and Genzel 1987). The density in the ionized filaments is less well known, but probably in the same range. For v = 100-400 km s^{-1}, the kinetic energy density is then $1/2\, \rho v^2 = 3 \times 10^{-5} - 5 \times 10^{-4}$ erg cm^{-3}. It appears that the kinetic energy density is at least an order of magnitude greater than the magnetic, so magnetic forces should not dominate the gas kinematics, although they may perturb it.

It should be noted that several streamers (notably the "eastern arm") have kinematics that do not fit with the rotating disk and morphologies very suggestive of solar flares. These filaments may well be strongly affected by magnetic fields.

Gaseous Drag

It is not clear how important hydrodynamic effects are in the gas kinematics. Particularly for the gas with Doppler shifts characteristic of circular orbits in a disk, drag forces seem unlikely to play a dominant role in the kinematics, but they may noticeably influence the morphology.

In the case of the ionized tail of the red supergiant IRS 7, it seems rather clear that hydrodynamic interaction with a wind from the center has a significant effect. The wind may also affect the "eastern arm" filaments which appear not to be in the molecular/ionized disk plane, and from their surface brightnesses must have lower density than the northern arm.

Although the nearly circular motion of the gas in the western-arc - northern-arm spiral suggests that it is not strongly affected by drag forces, Quinn and Sussman (1985) succeeded in making a model including drag which fits the observations. In their model (which was published before the spiral was apparent to the observers), the spiral results from a combination of tidal shearing and gaseous drag acting on an initially spherical cloud in the molecular ring. If the appropriate initial conditions are chosen (see Lacy *et al.* 1991) and the cloud is observed at the right time, it can fit both the morphology and the kinematics of the spiral seen in [Ne II]. However, there seems to be several problems with this model. First, a rather dense medium, which must have a density rising very steeply toward the center, is required to produce sufficient drag. Second, it would seem that the medium should quickly gain angular momentum for the infalling gas, and so no longer be able to slow the gas. And finally, a very large mass inflow rate is implied, particularly when the presence of ~ 300 $M_\odot$ of atomic gas, apprently moving with the ionized gas, is considered.

Lacy *et al.* (1991) prefer a model in which the spiral is a wave phenomenon in a disk with predominantly circular orbits. A large drag force is not then needed, but smaller perturbing forces are required to gather the gas into a spiral pattern. In a galactic disk, the non-radial gravitational forces due to the spiral wave itself maintains the spiral. The gas disk in the Galactic center is probably not sufficiently massive for this mechanism to work, but magnetic forces may serve instead.

A Black Hole?

The possibility of a black hole at the Galactic center was suggested at least 20 year ago (Lynden-Bell and Rees 1971). The kinematic evidence for a massive object is now rather convincing. However, the question remains: shouldn't we see a more prominent source at the center?

Two recent papers suggest that what we see is just what we should see. Melia (1992) calculated the spectrum expected if the wind from the IRS 16 cluster accretes spherically onto a ~ 10^6 $M_\odot$ black hole. He found that he could fit the spectrum of Sgr A* from 10^8 to 10^{19} Hz, and, in particular, predicted it should be undetectable in the infrared. The only discrepancy with the kinematic data is that he found a better fit with 1 x 10^6 $M_\odot$ than 2 x 10^6 $M_\odot$.

Mineshiga and Shields (1990) instead calculated the emission from an accretion disk. They found that a thermal instability in a disk can result in a lower luminosity, consistent with the observations of Sgr A*, 90 percent of the time, and a Seyfert-like luminosity for perhaps 100 years out of every 1000. An average accretion rate of as much as ~ 10^2 $M_\odot$ yr^{-1} is allowed.

Although we are still far from showing that 2 x 10^6 $M_\odot$ is contained within a Schwarzschild radius (~ 1 $R_\odot$), the evidence has probably reached the point where it would be more surprising if a black hole turned out not to exist than if one exists. Perhaps we will find out for sure if a flaring like that predicted by Mineshiga and Shields occurs. Observations of stellar velocities approaching 1000 km s^{-1} would also convince almost all skeptics.

Besides the question of a massive black hole, we are continuing to sort out the complex morphology and population of the region. Models have been suggested to explain perhaps half of the gas seen. It would be very desirable to have models for such features as the eastern ionized arm and the various anomalous molecular gas velocities. The distribution of red giants may simply trace the distribution of stellar mass. The origin of the central cluster of blue stars is unknown, but recent star formation may be required. Undoubtedly, the Galactic center will continue to be an interesting region for observational and theoretical study for some time to come.

REFERENCES

Aitken, D. K. *et al.* 1991, *Ap.J.*, **380**, 419.
Allen, D. A., Hyland, A. R., and Hillier, D. J. 1990, *MNRAS*, **244**, 706.
Allen, D. A. *et al.* 1983, *MNRAS*, **204**, 1145.
Backer, D. 1987, Symposium on the Galactic Center in Honor of Charles H. Townes, New York: Am. Inst. Phys.
Becklin, E. E. , and Neugebauer, G. 1968, *Ap.J.*, **151**, 145
Becklin, E. E., Gatley, I., and Werner, M. W. 1982, *Ap. J.*, **258**, 135.
Binney, J. *et al.* 1991, *MNRAS*, **252**, 210.
Brown, R. L., and Liszt, H. S. 1984, *ARAA*, **22**m 223.
Burton, M., and Allen, D. 1992, *Proc. Astr. Soc. Austr.*, in press.
DePoy, D. L., and Sharp, N. A., 1991, *A.J.*, **101**, 1324.
DePoy, D. L. *et al.* 1989, IAU #136, 411.
Eckart, A. *et al.* 1992, *Nature*, **355**, 526.
Gatley, I., *et al.* 1989, in I.A.U. No. 136, 361.
Geballe, T. R. 1989, in I.A.U. No. 136, 469.
Genzel, R., and Townes, C. H. 1987, *ARAA*, **25**, 377.
Glass, I. S., Catchpole, R. M., and Whitelock, P. A. 1987, *MNRAS*, **227**, 373.
Hildebrand, K. H. *et al.* 1990, *Ap.J.*, **362**, 114.
Jackson, J. M. *et al.* 1992, preprint.
Killeen, N. E. B. *et al.* 1990, IAU #140, 382.

Killeen, N. E. B., and Lo, K. Y. 1989, IAU #136, 453.
Krabbe, A. *et al.* 1991, *Ap.J.*, **382**, L19.
Lacy, J. H., Achtermann, J. M., and Serabyn, E. 1991, *Ap.J.*, **380**, L71.
Lindqvist, M. *et al.* 1989, IAU #136, 503.
Lo, K. Y. 1989, IAU #136, 527.
Lynden-bell, D., and Rees, M. J. 1971, *MNRAS*, **152**, 461.
Melia, F. 1992, *Ap.J.*, **387**, L25.
Mineshiga, S., and Shields, G. A. 1990, *Ap.J.*, **351**, 47.
Morris, M. 1989, *The Center of the Galaxy*, I.A.U. Symp. No. 136, Dordrecht: Kluwer Academic Publisher.
Morris, M. 1990, IAU #140, 361.
Oort, J. H. 1977, *ARAA*, **15**, 295.
Pedlar, A. *et al.* 1989, *Ap.J.*, **342**, 769.
Quinn, P. J., and Sussman, F. J. 1985, *Ap.J.*, **288**, 377.
Riegler, G. R., and Blandford, R. D. 1982, *The Galactic Center*, AIP Conf. Proc. No. 83, New York: Am. Inst. Phys.
Rieke, G. H., and Lebofsky, M. J. 1987, AIP # 155, 91.
Rieke, G. H., and Rieke, M. J. 1988, *Ap. J.*, **330**, L33.
Roberts, D. A. *et al.* 1991, *Ap.J.*, **366**, L15.
Roberts, D. A., and Goss, W. M. 1992, *Ap. J. Suppl.*, in press.
Schwarz, U. J., Bregman, J. D., and van Gorkham, J. H. 1989, *A&A*, **215**, 33.
Schwarz, U. J., and Lasenby, J. 1990, IAU #140
Sellgren, K. 1989, IAU #136, 1477.
Sellgren, K. *et al.* 1990, *Ap.J.*, **359**, 112.
Serabyn, E. 1989, IAU #136, 437
Serabyn, E., Lacy, J. H., and Achtermann, J. M. 1991, *Ap.J.*, **378**, 557.
Smith, C. H., Aitken, D. K., and Roche, P. F. 1990, *MNRAS*, **246**, 1.
Sutton, E. C., *et al.* 1990, *Ap. J.*, **348**, 503.
Yusef-Zaden, F., and Melia, F. 1992, *Ap.J.*, **385**, L41.
Yusef-Zadeh, F., and Morris, M. 1991, *Ap.J.*, **371**, L59.
Yusef-Zadeh, F. *et al.* 1989, IAU #136, 443.

Note added in proof: Zylka, R., Mezger, P. G., and Lesch, H. 1992, *A&A*, **261**, 119, report detection of Sgr A* at 350, 870, and 1300 μm wavelength. The sub-millimeter spectrum can be modelled either by a dust disk or a self-absorbed synchrotron source. They suggest that a black hole of 1-2 x 10^6 $M_\odot$, accreting $\dot{M} \sim 10^{-6}$ $M_\odot$ yr^{-1} could explain all observations of Sgr A* from 1 μm to centimeter wavelengths. They also mapped the 870 μm emission which closely follows the one-armed spiral seen in the ionized gas (Lacy *et al.* 1991). Their apparent detection of an accretion disk, particularly if confirmed at shorter wavelengths, will provide strong evidence for a central massive black hole.

The Minnesota Lectures on Clusters of Galaxies and Large-Scale Structure
ASP Conference Series, Vol. 39, 1993
Roberta M Humphreys (ed.)

Gas as a Tracer of Galactic Structure

John M. Dickey
Department of Astronomy
University of Minnesota

Introduction

The interstellar obscuration at optical wavelengths blocks our view of stars over most of the area of the Galactic plane, so that it is only by recourse to other wavelength bands that we can survey the Galactic disk on a large scale. As anticipated by Oort in the early 1940's, radio spectroscopy has become invaluable as a technique for mapping the structure and motions of the Milky Way disk. Only a few years after the first discovery of the 21-cm line of atomic hydrogen, observers at Leiden and Sydney had measured the rotation curve of the inner Galaxy, and made a rough map of the distribution of gas (van de Hulst, Muller and Oort, 1954, see Sullivan, 1984 for a historical review). Many surveys have been made since then, both in the 21 cm line, in recombination lines which trace ionized hydrogen, and in several molecular species which trace the dense, cold molecular component. Although the gas serves as our primary tracer of kinematics in the Milky Way and other galaxies, from which we can derive the gravitational force and hence the distribution of total mass, the gas itself is dynamically insignificant, as it constitutes only 10% or less of the total mass in most regions of most spiral galaxies. In this chapter I summarize what we learn about the structure of the Milky Way from observations of centimeter and millimeter wavelength lines from the interstellar gas.

I. Phases of the Gas

The interstellar medium is filled with gas of various temperatures and densities. There is rough energy and pressure balance between different regions with different properties, so we speak of these as different "phases" of the interstellar medium. The analogy to ice, water and steam is apt, since all the interstellar phases have the same elemental compositions, yet differ in temperature and density by factors greater than the differences between solid, liquid and gas on the earth. Representative values for the physical properties of the interstellar phases are summarized in the table. Many of these quantities are controversial, in particular the filling factors, which change with height above the plane z. Values in the table are representative for $z = 0$, at high z the filling factors of the warm neutral medium and particularly the warm ionized medium increase to perhaps 40% and 30% respectively, at the expense of the cool neutral medium in the diffuse clouds, which are very rare above $z \approx 200$ pc. (An excellent review of the thermal phases of the interstellar gas and their filling factors is given by Kulkarni and Heiles, 1988.)

Table : The Interstellar Phases

name	temperature	density	filling factor
hot ionized medium	10^6 K	10^{-3} cm^{-3}	~50 %
warm ionized medium	7500 K	0.3 cm^{-3}	~10 %
warm neutral medium	6000 K	0.3 cm^{-3}	~35 %
diffuse clouds	60 K	30 to 100 cm^{-3}	~ 5 %
molecular clouds	10 K	> 10^3 cm^{-3}	< 1 %

There is rough pressure equilibrium among the phases, with P/k in the range 10^3 to 10^4 K cm^{-3}, except for the molecular clouds, which are partially confined by self-gravity. This was a critical feature of the Field, Goldsmith and Habing model (1969, see review by Dalgarno and McCray, 1972), but it is less critical in more modern theories of the thermodynamics of the gas, like that of McKee and Ostriker (1977) and Heiles (1990). On large scales, the interstellar pressure must be in equilibrium with the gravitational potential depth or else the gas will either escape or collapse. This must determine which phases are viable in different environments. But precise pressure equilibrium among the phases is not expected, since the sound crossing time for interstellar clouds is typically millions of years, which is of the same order as the typical interval between being swept over by supernova remnants.

Tracing the Molecular Gas

Although the molecular gas has a very small volume filling factor, it contains a large fraction of the mass of the interstellar medium. Unfortunately, the dominant molecular species, H_2, has no transitions at accessible frequencies which are normally excited at the low temperatures of most molecular clouds. So to trace the molecular gas we rely on observations of molecules like CO, HCN, H_2CO, C_3H_2, CH, OH, and CS, which constitute 10^{-4} of the number of molecules, or less. The lowest rotational transition of CO, at wavelength 2.6 mm (115.271 GHz), is one of the most convenient to observe. The brightness of this line depends on the population of the upper state, which is in equilibrium with the kinetic temperature as long as the collisional excitation rate is comparable to or greater than the spontaneous de-excitation rate, A_{21}. In molecular clouds this is the case when the density is greater than about 10^3 cm^{-3}. Another useful tracer, the CS molecule with several transitions at wavelengths between 2 and 4 mm, traces densities of 10^5 cm^{-3} and higher. We find its emission concentrated much more strongly toward the inner Galaxy than the CO emission.

The CO line is optically thick in most molecular clouds, so to trace the optical depth and the column density of gas we observe the corresponding line of the ^{13}C isotope, ^{13}CO, at 110.201 GHz, which is more often optically thin, although not always. It turns out that the velocity integrals of the emission spectra in ^{13}CO and ^{12}CO are fairly well correlated, so in spite of the fact that the ^{12}CO line is optically thick, it is used roughly as a tracer of molecular column density and hence of molecular mass. (This empirical result has a theoretical basis in the temperature and velocity structure of clouds, see reviews by Combes, 1991, and Turner, 1988). The conversion factor between the observed brightness temperature of the 115 GHz CO line, T_R^*, integrated over all velocities in the spectrum, and the column density of H_2, $N(H_2)$, is

$$\frac{N(H_2)}{\mathrm{cm}^{-2}} = (1 \text{ to } 4) \times 10^{20} \frac{\int T_R^* \, dv}{\mathrm{K\ km\ s^{-1}}} \qquad 1.$$

where the range, 1 to 4, corresponds to a range of physical conditions, with higher values applying to colder, more optically thick environments.

Tracing the Atomic Gas

Atomic hydrogen dominates in two interstellar phases, the warm neutral medium ("WNM") and the diffuse clouds, which are often referred to as cool neutral medium ("CNM"), but which actually contain a mixture of roughly equal parts WNM and CNM. The 21-cm line in emission traces the column density of the atomic gas at all temperatures, as the emission coefficient is independent of temperature, so

$$\frac{N(\mathrm{H\ I})}{\mathrm{cm}^{-2}} = 1.8 \times 10^{18} \frac{\int T_B \, dv}{\mathrm{K\ km\ s^{-1}}} \qquad 2.$$

where T_B is the brightness temperature of the 21-cm line. When the line is optically thick this must be corrected for absorption, but this correction is no more than 30% in most emission surveys of the Milky Way. The CNM can be traced separately from the WNM by observing absorption by the 21-cm line, as the opacity of the line depends on the inverse of the temperature, so that only cold gas ($T < 300$ K, roughly) contributes to the absorption.

The warm ionized gas ("WIM") can be traced by free-free radio continuum emission and by recombination lines. In the radio these lines correspond to very high quantum numbers, n; transitions as high as the $n = 630 \rightarrow 629$ have been detected. The frequencies of recombination lines for hydrogenic species (single electron) are given by the Rydberg formula :

$$\frac{\nu}{\mathrm{GHz}} = \frac{E_{n+\Delta n \rightarrow n}}{h} \qquad 3.$$

$$= 3.2898 \times 10^6 \; Z^2 \left(1 - \frac{5.486 \times 10^{-4}}{M_{AMU}}\right) \left[\frac{1}{n^2} - \frac{1}{(n+\Delta n)^2}\right]$$

where M_{AMU} is the mass of the nucleus in atomic mass units (M for hydrogen = 1.0078), and Z is the nuclear charge (Z = 1 for hydrogen). The emission coefficient for recombination lines is proportional to density squared, so their brightness is dominated by regions of high density of ionized gas, particularly H II regions. The average density of electrons in the interstellar medium as a whole is given better by pulsar dispersion measures, which give a value of $< n_e > \approx 0.025$ to 0.03 cm^{-3} averaged over volume. The pulsar dispersion also gives a scale height (ie. the e-folding height above the plane) for the warm ionized medium of about 1 kpc (reviewed recently by Nordgren, Cordes and Terzian, 1992)

The Hot Gas

The hot phase of the interstellar medium was first detected in the early 1970's in soft x-ray emission and in ultraviolet absorption lines such as O VI and Fe XXV. The temperature of this phase varies from several x 10^5 K up to 10^8 K or more. Energy is continuously being injected into this phase by supernova remnants and stellar winds, which can sweep out bubbles of hot gas as large as tens of parsecs or more. Clustered supernovae in the disk can drive tunnels or chimneys of hot gas upward into the lower halo, where the gas eventually cools by soft x-ray radiation and condenses into clouds of WIM and WNM, which fall back into the plane in a "galactic fountain" process. This is the origin of some forms of high velocity clouds (HVC's) seen in the 21-cm line. For reviews of the theory and observations of this phase see Spitzer (1990) and McCammon and Sanders (1990).

II. Vertical Distribution of the Gas

The Force Law

In the solar neighborhood, observations of stellar densities and velocities give us a way to measure the gravitational force in the z direction (perpendicular to the Galactic plane) as a function of z. The gravitational acceleration as a function of z we term K_z, analogous to g on earth. A rough

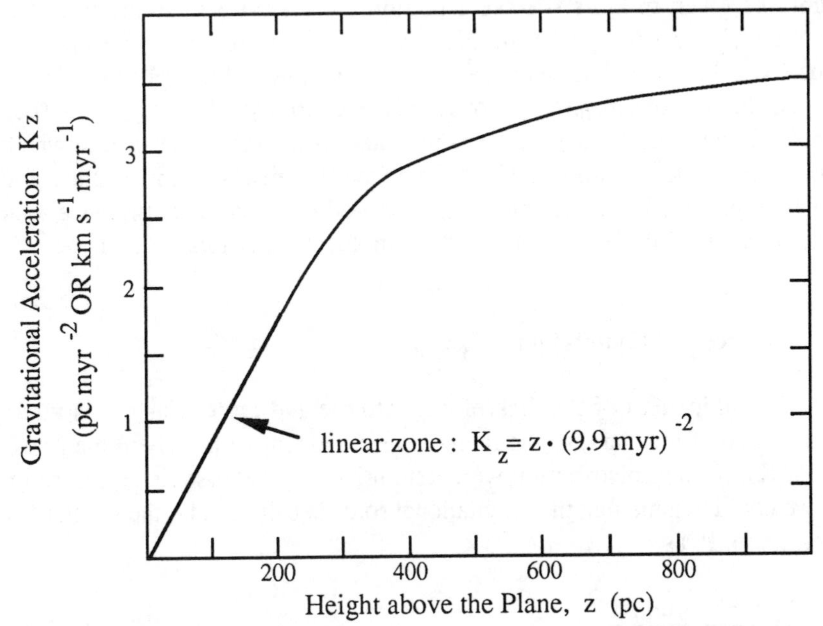

Figure 1 : The gravitational acceleration perpendicular to the plane, K_z.

sketch of K_z as a function of z is shown on figure 1. The linear portion corresponds to constant density, since by Gauss's law in one dimension, the divergence of the gravitational potential Φ is

$$\frac{d^2\Phi}{dz^2} = 4\pi G \rho \qquad 4.$$

where ρ is the total mass density (of which the gas is a small fraction) and G is Newton's constant. Since K_z is the gradient of Φ

$$K_z = -\frac{d\Phi}{dz} = -4\pi G \int_0^z \rho(z)\,dz = -\omega^2 z \cong \frac{-z}{(9.9\ \text{myr})^2} \qquad 5.$$

where the latter two equations apply in the region of roughly constant density $\rho(0)$ below $z \approx 300$ pc. The slope of K_z in this linear zone, ω, tells us immediately the local **total** mass density, $\rho(0)$, first measured by Oort (1932) and now termed the "Oort Limit" .

A convenience of "Galactic units", i.e. pc for distance, millions of years (myr) for time, and solar masses ($M_\odot$) for mass, is that velocities can roughly be given in km s^{-1}, since 1 pc myr^{-1} = 1 km s^{-1} to about 2%. The unit of density is 1 $M_\odot pc^{-3}$ = 43 H atoms cm^{-3}, and for column density 1 $M_\odot pc^{-2} = 1.3 \times 10^{20}$ H atoms cm^{-2}. G has the value 4.3×10^{-3} (km $s^{-1})^2$ pc $M_\odot^{-1}$, so that the observed value of $\omega = (9.9 \pm 1.6\ \text{myr})^{-1}$ gives $\rho(0) \cong 0.19$ $M_\odot pc^{-3}$. Integrating K_z over z gives the total mass surface density, Σ, which is 50 to 75 $M_\odot pc^{-2}$ within $|z| < 700$ pc (see the discussion in Binney and Tremaine, p. 200 ff). In the solar neighborhood the atomic gas shows a total surface density of about 5 $M_\odot pc^{-2}$ and the molecular gas about 1.2 $M_\odot pc^{-2}$.

Hydrostatic Equilibrium

The thickness of the disk of cool atomic and molecular gas is much less than that of the stars, so the interstellar clouds move mostly in the linear zone of K_z. If their distribution were determined by hydrostatic equilibrium, then we could assume that the gravitational force is balanced by the gradient of the pressure, P, as

$$\frac{1}{\rho_g}\frac{dP}{dz} = -\frac{d\Phi}{dz} = K_z \qquad 6.$$

where ρ_g is the density of the gas. Such a force balance does exist, but we must generalize the pressure to include more than just the microscopic kinetic motions of atoms and molecules. In particular, the pressure due to bulk motions of the clouds, plus the magnetic field pressure and the cosmic ray pressure which couples through the magnetic field, turn out to be larger than the kinetic pressure (see the discussion of Spitzer, 1977, chap. 11). Imagining for the moment that all these contributions to the pressure can be expressed as an effective velocity squared, $< v^2 >$, where the total pressure is $P = \rho_g < v^2 >$, then

$$\frac{d}{dz}(<v^2> \rho_g) = \rho_g \frac{d\Phi}{dz} \quad 7.$$

is a restatement of hydrostatic equilibrium. If the gas is effectively "isothermal", so that $<v^2>$ is independent of z, then this simplifies to

$$<v^2> \frac{d \ln \rho_g}{dz} = \frac{d\Phi}{dz} \quad 8.$$

which we integrate in the "linear zone" of K_z (below about 300 pc in z), giving for the gas density

$$\rho_g(z) = \rho_g(0) \exp\left(\frac{-z^2}{2 \frac{<v^2>}{\omega^2}} \right) \quad 9.$$

that is, a Gaussian density distribution with dispersion $\sigma_z = \sqrt{<v^2>} / \omega$, with $\omega = \sqrt{4 \pi G \rho(0)} = (9.9 \pm 1.6 \text{ myr})^{-1}$ as above. This leads to a relationship between the surface density of gas, $\Sigma_g = \sqrt{2\pi}\, \sigma_z\, \rho_g(0)$, and the effective pressure, $P = \rho_g v^2$:

$$P = \sigma_z\, \omega^2\, \Sigma_g\, (2\pi)^{-\frac{1}{2}} \quad 10.$$

which is useful in cases such as face-on spirals where we cannot measure the space density directly, but we can observe the surface density and make reasonable guesses for ω and σ_z. In Galactic units this quantity has dimensions $M_\odot\, pc^{-3}\, (km\, s^{-1})^2$, which converts to the more familiar P/k units of K cm^{-3} by multiplying by 5200 (the z motions are a one dimensional problem so $mv^2 = kT$); thus the molecular clouds in the solar neighborhood would give $P \cong 0.6\, M_\odot pc^{-3}\, (km\, s^{-1})^2$ or 3000 K cm^{-3}.

Observations show the scale heights, σ_z, of the cool gas phases to be about 120 pc for the diffuse clouds and about 75 pc for the molecular clouds. Using these we can deduce the effective rms velocities of the gas in these two phases directly from eq. 9. We find $\sqrt{<v^2>} \cong 11$ km s^{-1} and 7 km s^{-1} respectively. The observed rms cloud velocities are about 5 to 6 km s^{-1}, which shows that other contributions to the effective pressure of the atomic gas are significant **and/or** that the thermodynamic environment changes with

z. If the heating and cooling rates change with z, e.g. because of changes in the ambient radiation field, this may cause the cloud phase to be no longer in stable thermodynamic equilibrium. These complications are more serious for the atomic and ionized media and perhaps for the smaller molecular clouds. The giant molecular clouds (GMC's) have such high masses (10^4 to 10^6 $M_\odot$) that they move ballistically in the gravitational field, and so for them it is reasonable to equate $\sqrt{< v^2 >}$ to $\omega\ \sigma_z$.

The warm neutral and ionized media reach much higher in z (see Lockman and Gehman, 1991), so that their densites as functions of z would be Gaussians at low z and approach exponentials at z >> 300 pc, if they were in hydrostatic equilibrium. In fact the magnetic field and cosmic ray contributions to the pressure in these phases are of the same order as the microscopic gas motions, and there is evidence for large scale bulk motions with supersonic speeds, so the effective $< v^2 >$ may be a complicated function of z. For the atomic gas as a whole the scale height is about 180 pc and the surface density is about 5 $M_\odot pc^{-2}$ in the solar neighborhood, so equation 10 gives $P \cong 3\ M_\odot pc^{-3}\ (km\ s^{-1})^2 = 1.6 \times 10^4\ cm^{-3}$ K. This is much more than the gas pressure alone, so evidently the effective pressure in this phase is supplemented by these other factors.

III. Motions in the Galactic Plane

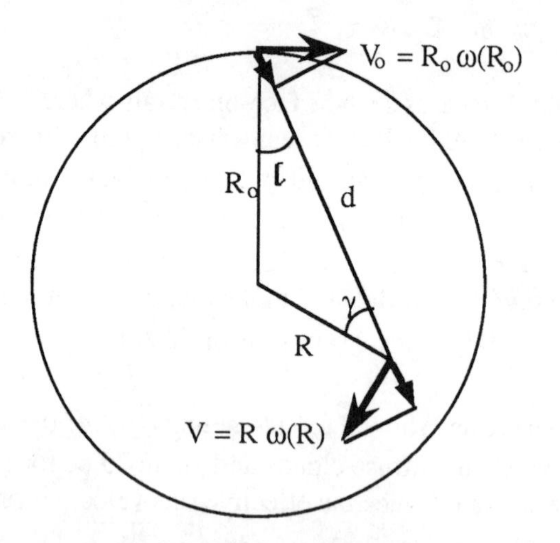

Figure 2 : Geometry for determining radial velocities in the Galactic plane.

Radial Velocities in a Differentially Rotating Disk

Spectroscopy of the interstellar gas gives us most of the information that we have about the large scale motions of the Milky Way disk. The basic geometry used to derive the projected radial velocity resulting from simple circular differential rotation is shown on figure 2. This is a face-on view of the Galaxy, with the solar circle (circular orbit at the sun's distance from the center, $R_\odot$) drawn, and a sample element of gas at a distance d from the sun, which corresponds to galactocentric radius R and longitude ℓ. The radial velocity, v_r, which we would measure for this gas (in the Local Standard of Rest frame, which is a good approximation to a circular orbit at $R_\odot$) is the difference between its velocity projected onto the line of sight minus the projection of the sun's velocity, i.e.

$$v_r(R,\ell) = R\,\omega(R)\cos(90-\gamma) - R_\odot\omega(R_\odot)\cos(90-\ell) \qquad 11.$$

where $\omega(R)$ is the rotation curve in angular velocity units, (radians per myr) *or* (km s^{-1} pc^{-1}), as a function of galactocentric radius R. Using trigonometric identities and the law of sines, i.e. $R\sin\gamma = R_\odot \sin\ell$, gives the simple and very useful formula :

$$v_r(R,\ell) = R_\odot \sin(\ell)\,[\,\omega(R) - \omega(R_\odot)\,]\cos(b) \qquad 12.$$

where the cos(b) term generalizes eq. 11 for points above or below the plane, assuming cylindrical rotation. Using the law of cosines, i.e. $R^2 = R_\odot^2 + r^2 - 2\,r\,R_\odot\cos(\ell)$, we can convert $v_r(R,\ell)$ in equation 11 to $v_r(r,\ell)$. The **tangent point,** or **sub-central point**, is the point on the line of sight closest to the Galactic center, where the line of sight is tangent to the circular orbit of the gas. On lines of sight through the inner Galaxy ($-90^o < \ell < 90^o$), for any rotation curve, the function $v(r,\ell)$ has the same value at two different points equally spaced on opposite sides of the tangent point. Thus inverting $v_r(r,\ell)$ to get $r(v_r,\ell)$ is ambiguous inside the solar circle. The velocity of the tangent point is an extremum of the velocities along its line of sight (as long as $\omega(R)$ is not a pathological function of R). This is called the **terminal velocity**. The locus of these tangent points defines a circle centered half way between the sun and the Galactic center. Measuring spectra throughout either the first quadrant ($0^o < \ell < 90^o$) or the fourth quadrant ($270^o < \ell < 360^o$) allows us to measure the terminal velocity as a function of R, and so determine the rotation curve $\omega(R)$ (comparisons of results from CO and H I are given by Clemens, 1985 and Brand and Blitz, 1993).

The Oort Constants

For the special case of nearby objects, $r \ll R_\odot$ we can make a Taylor expansion for $\omega(R) - \omega(R_\odot)$ and simplify equation 12. Thus

$$\omega(R) - \omega(R_\odot) \cong \left.\frac{d\omega}{dR}\right]_{R_o} \Delta R \cong \left.\frac{d\omega}{dR}\right]_{R_o} [-r\cos(\ell)] \qquad 13.$$

where the derivatives are evaluated at the solar circle. Substituting this into equation 12 and using another trigonometric identity gives :

$$v_r(r,\ell) \cong r\sin(2\ell)\left(- \frac{R_\odot}{2}\left.\frac{d\omega}{dR}\right]_{R_o} \right) \qquad 14.$$

where the quantity in the brackets is a constant characteristic of the solar circle first determined by Oort called the A constant. A similar analysis to that of equations 11 - 13 for tangential velocities leads to

$$v_t(r,\ell) \cong r\,(\,A\cos(2\ell) + B\,) \qquad 15.$$

where B is the second Oort constant, defined by

$$B \equiv -\frac{1}{2}\left(\left.\frac{dV}{dR}\right]_{R_o} + \left.\frac{V}{R}\right]_{R_o} \right)$$
$$= -12 \pm 3 \text{ km s}^{-1}\text{ kpc}^{-1} \quad (\text{or } -0.012 \pm 0.003 \text{ km s}^{-1}\text{ pc}^{-1}) \qquad 16.$$

(numerical values are taken from Binney and Tremaine, 1987, p.17 and Kerr and Lynden-Bell, 1986). Note that A can be written in a very similar form

$$A \equiv -\frac{1}{2}\left(\left.\frac{dV}{dR}\right]_{R_o} - \left.\frac{V}{R}\right]_{R_o} \right) = \left(- \frac{R_\odot}{2}\left.\frac{d\omega}{dR}\right]_{R_o} \right)$$
$$= 14.5 \pm 1.3 \text{ km s}^{-1}\text{ kpc}^{-1} \qquad 17.$$

There are several identities involving A and B, some of the more useful (and obvious) are :

$$A - B = \frac{V_\odot}{R_\odot} = \omega_\odot = 26.4 \pm 1.9 \text{ km s}^{-1}\text{ kpc}^{-1} \qquad 18.$$

$$A + B = -\left.\frac{dV}{dR}\right]_{R_o} = 2.5 \pm 4 \text{ km s}^{-1} \text{ pc}^{-1} \quad 19.$$

$$-2 A R_\odot = \frac{dv_{max}}{d\sin(\ell)} = 120 \pm 10 \text{ km s}^{-1} \quad 20.$$

where in the third equation v_{max} is the terminal velocity, i.e. the extremum velocity measured along a line of sight in the inner Galaxy at galactic longitude ℓ. These values are consistent with the present IAU "standards" for the sun's galactocentric radius and the LSR orbital velocity : $R_\odot \equiv 8.5$ kpc and $V_\odot \equiv 220$ km s^{-1}, although there is evidence that these values are high by about 5% (Merrifield 1992).

III. The Velocity Gradient and the ℓ-v Diagram

Using observed velocities and equation 12 we can estimate distances to objects ("kinematic distances") in the outer Galaxy, and in the inner Galaxy also but for the distance ambiguity between the near and far sides of the sub-central point. A simple power law which fits the rotation curve data in the range 2 < R < 15 kpc was determined by Brand and Blitz (1993, see Burton, 1988 and Merrifield, 1992) :

$$\frac{V(R)}{V_o(R_o)} = \frac{R\,\omega(R)}{R_o\,\omega(R_o)} = 1.0074\left(\frac{R}{R_o}\right)^{0.0382} + 0.00698 \quad 21.$$

with $R_\odot$ and $v_\odot$ as defined above. (This does not give a value for A quite consistent with the locally measured quantity in equation 17, so it should not be used in the solar neighborhood.) Kinematic distances are limited in accuracy by random motions, and by local departures from circular rotation, which are typically 10% to 15%, but may be systematically greater in some directions as suggested in the non-circular model of Blitz and Spergel (1991).

Our best method of mapping the large scale structure of the disk of the Galaxy is by surveying spectral line emission from the interstellar gas, yet having done many such surveys we still do not have a clear picture of how the Galaxy would look seen face-on. This is not so much because of the distance ambiguity as because of the sensitivity of the velocity gradient to kinematic distortions of simple circular rotation. For an optically thin emission line for which the level populations are in Boltzmann equilibrium with $h\nu \ll kT$, the measured brightness at a given velocity is proportional to the column density of gas at that velocity in the telescope beam. The distance along the line of sight which contributes to a given velocity interval is proportional to the inverse of the slope of the velocity vs. distance relation, $v_r(r,\ell)$. This slope, $\frac{dv}{dr}$, called the **velocity gradient,** is a smooth function of position in the inner Galaxy, depending only on the shape of the rotation curve. In the solar neighborhood it is just A sin (2ℓ), and it typically has absolute value in the range 5 to 15 km s^{-1} kpc^{-1}, changing sign at the the tangent point. Systematic deviations from non-circular rotation of as little as 10 to 15 km s^{-1} can cause as much structure in line profiles as major density variations because they can double the velocity gradient locally or even cause it to change sign. Since we have no means of mapping out such relatively minor departures from circular rotation, it is difficult to translate from a compendium of spectra back to a face-on image of the Galaxy.

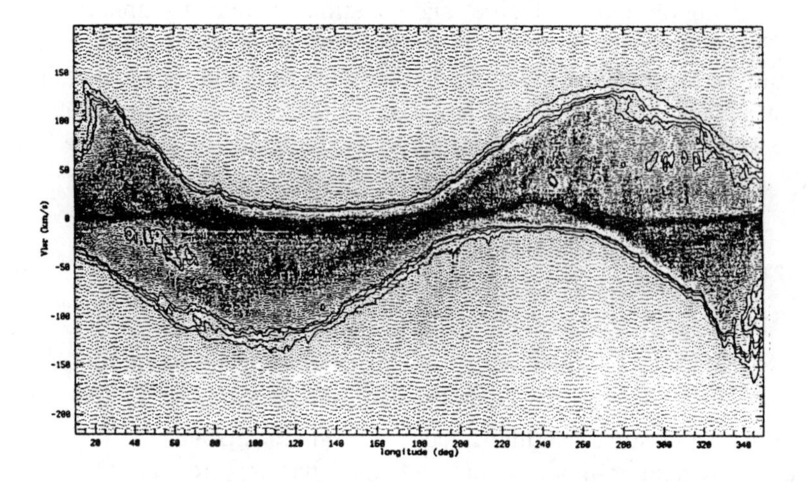

Figure 3 : The ℓ-v diagram for the 21-cm emission from neutral atomic hydrogen, taken from the data compilation of Blitz and Spergel (1991). The structure at low velocities and in the inner galaxy is suppressed by the grey scale representation.

Since we cannot map the observations back onto an image of the Galaxy, the best presentation of survey results is on the plane of longitude vs. radial velocity as measured, giving an "ℓ-v diagram", i.e. a contour map of brightness (or brightness temperature) as a function of ℓ and v_r. Examples of these are shown in figures 3 and 4 for H I and CO surveys. Structures on these figures corresponding to kinematic and density inhomogeneities on large scales are apparent. The most striking feature on the CO surveys is the "molecular ring", a very high concentration of molecular gas at $3 < R < 4$ kpc. Neither of these tracers is optically thin, so the highest brightness temperatures in both cases may be saturated, i.e. the column density is greater than what would be predicted by a linear conversion from the measured brightness.

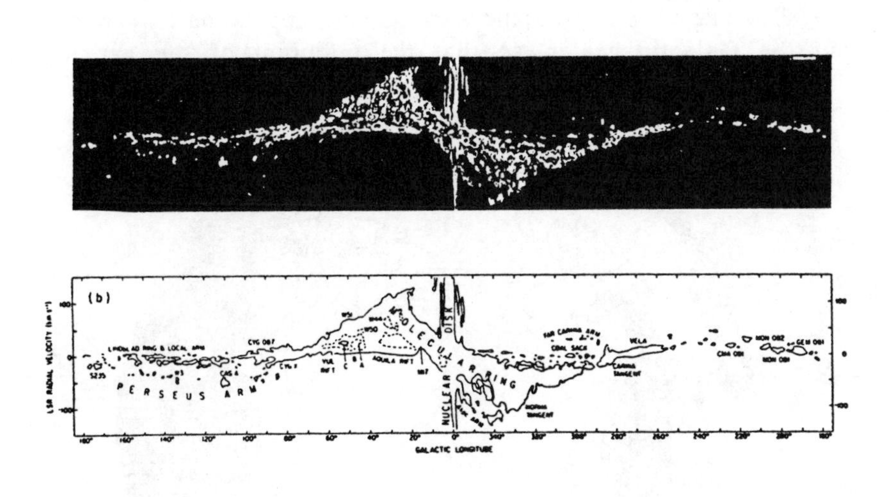

Figure 4 : The ℓ-v diagram of CO emission, with a guide to some commonly discussed features, taken from Dame et al. (1987).

IV. The Shape of the Gas Disk

The distribution of the gas in the inner Milky Way disk is very thin and very flat. Departures of the center of the gas distribution in z from the midplane of the Galaxy are less than 50 pc wherever they can be measured in the inner Galaxy, both for CO and for H I. The half width of the molecular layer is only 75 pc or so, whereas its radial scale length is ~4 kpc, so the thickness is only 1% of the radius. The z-thickness of the atomic hydrogen is about twice that of the CO, but its radial distribution is larger also, reaching to about 20 kpc radius. The surface density of the atomic gas is roughly constant at ~5 $M_\odot$ pc^{-2} in the range 3 < R < 18 kpc, whereas the molecular gas has a roughly exponential decrease with radius. The peak surface density of the molecular gas in the molecular ring is about 8 $M_\odot$ pc^{-2} at R ≅ 4.5 kpc. There is a hole in the inner disk, for 1 < R < 2.5 kpc both the atomic and molecular gas decrease. Inside of 1 kpc there is a molecular disk whose surface density is as high as 60 $M_\odot$ pc^{-2}.

Outside the solar circle the atomic gas dominates the Galactic disk, whereas in the inner Galaxy the molecular gas has a higher surface density, since its distribution is exponential and the atomic gas surface density is roughly constant. In the outer Galaxy the disk becomes less thin and flat. The scale height of the atomic gas flares up to as much as 1 kpc in some places. The plane has a systematic warp which is approximately symmetric about the Galactic center, so that the midpoint of the gas layer

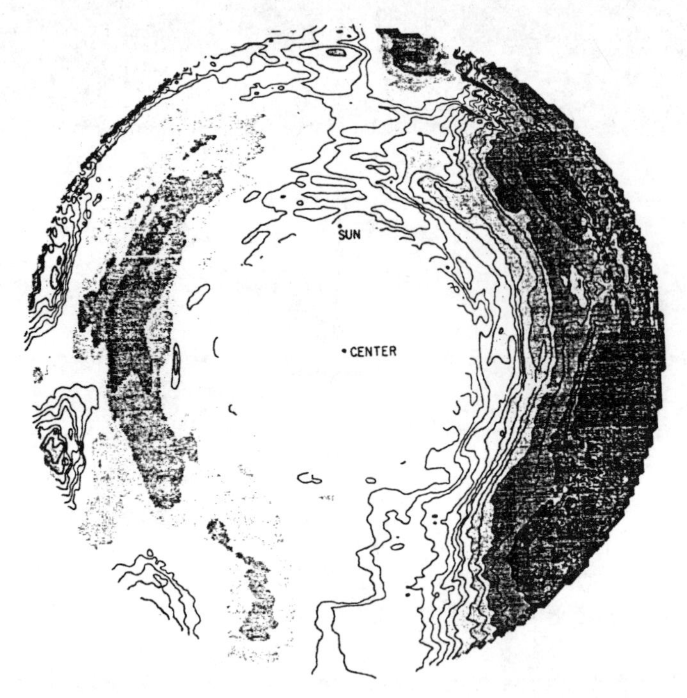

Figure 5 : The warp of the outer Galaxy disk, from Burton (1988).

extends as far as a kiloparsec from the plane of the inner Galaxy, as shown on figure 5. The line of nodes is close to the sun - center line, and quite straight.

V. Unanswered Questions

Many questions both large and small remain unanswered about the distribution of gas in the Galaxy. Some of these will become more tractable with new telescopes which are under development, others must await a breakthrough of creativity to find new ways to use old instruments, or to reinterpret existing data. Here I mention a few areas where I expect progress in the next few years will be particularly fruitful.

Small Scale Structure

We often use the word "cloud" loosely to describe structures in the interstellar gas with low filling factor. The implied analogy to clouds in the earth's atmosphere may be very misleading. In fact there is a hierarchy of structure on a range of scales from the giant molecular clouds with sizes of 30 to perhaps 100 pc and masses of up to 10^6 $M_\odot$ down to turbulence which is traced by pulsar scintillations with scales of a few AU. Presumably this hierarchy represents some kind of spectrum, but how a given volume of the interstellar gas evolves from one level to another, and what physical principles determine the characteristics of the structure we do not know. The IRAS survey of infrared "cirrus" has been very useful in providing data with high dynamic range in column density and spatial coverage (Houlahan and Scalo, 1991).

The Halo and Fountain

The dynamics of the disks of the Milky Way and other spiral galaxies suggest extended dark halos of matter. In addition, the abundance of absorption lines toward quasars suggests that galaxies have, or at least had in the past, extended halos of gas including some heavy elements. Observations of uv-absorption at high latitudes show a population of clouds with low column densities and a very high scale height. Some theories suggest a "fountain" process, driven by supernova explosions, which continuously injects matter into the halo from the disk, and which could enrich the outer disk with processed material. A fountain is not the only possibility, if the supernova rate were high enough then a wind would be driven out of the disk faster than the escape velocity from the halo (Yamauchi et al. 1990). If the supernova rate were very low then the filling factor of the hot ionized medium would shrink to only a small fraction of the total. The Milky Way seems to be in the intermediate range, where a fountain results, but how peculiar this is compared to other galaxies we do not know for sure (see Norman and Ikeuchi, 1989). Questions of the "disk-halo connection", and particularly how the mixture of thermal phases changes with height above the plane, are currently the topics of much active research.

The Magnetic Field

On small scales the magnetic field has sufficient pressure and energy density to dominate the dynamics of the interstellar gas. On large scales the magnetic field can be gradually amplified by differential rotation of the disk (reviewed by Zweibel, 1987). In the z direction the disk is unstable to the magnetic Rayleigh-Taylor instability (the "Parker Instability", Parker, 1966) so that if the pressure due to cosmic rays and magnetic fields becomes high enough this "light fluid" can lift itself out of the disk, separating from the clouds which are gravitationally bound. On small scales the magnetic field has irregularities which may result from a spectrum of Alfven waves, related to the turbulence spectrum which describes the small scale density and velocity structure. How the magnetic field in the disk comes into equilibrium with differential rotation on large scales, and how the dynamics of the gas are governed by the magnetic field on small scales are both extremely interesting issues.

Formation and Evolution

The structure of the Milky Way holds tantalizing clues to the process of galaxy formation. Large scale variations in the abundance of heavy elements in the gas as well as the stars suggest a non-uniform star formation history. The Milky Way has evidently accreted some primordial matter since its formation, and/or lost enriched matter to a wind. The links between the distributions of atomic and molecular gas and the past and current star formation rates are understood only vaguely. The distribution of cloud random velocities is apparently determined by the star formation rate, but we do not know whether the cloud ensemble relaxes kinematically through collisions, nor what determines the rms gas velocity in the outer disk where star formation has not "turned on". Long term secular trends in the disk, like mass inflow or radial progression of the molecular cloud/star formation zone, are not well understood. We may hope for progress on all these questions which will enhance our understanding of the galaxy as we see it today.

This chapter follows in part the structure of two lectures given by Leo Blitz during this visitors' course. I am grateful to him for permission to use his material, and for helpful comments.

References

Binney, J. and Tremaine, S., 1987, **Galactic Dynamics**, (Princeton: Princeton University Press).

Blitz, L. and Spergel, D.N., 1991, Ap.J., 370, 205.

Brand, J. and Blitz, L., 1993, Astron. Astrop. in press.

Burton, W.B., 1988, in *Galactic and Extragalactic Radio Astronomy*, 2nd ed., eds. G.L. Verschuur and K. Kellerman, (New York : Springer-Verlag) p. 295.

Clemens, D.P., 1985, Ap. J. 295, 422.

Combes, F., 1991, Ann. Rev. Astr. Astrop. 29, 195.

Dalgarno, A., and McCray, R.A., 1972, A.R.A.A. 10, 375.

Dame, T.M., Ungerechts, H., Cohen, R.S., de Geus, E.J., Grenier, I.A., et al., 1987, *Ap. J.*, **322**, 706.

Field, G.B., Goldsmith, D.W., and Habing, H.J., 1969, Ap. J. Lett. 155, L149.

Heiles, C., 1990, Ap. J. 354, 483.

Houlahan, P. and Scalo, J.M., 1990, *Ap. J. Suppl.*, **72**, 133.

Kerr, F.J., and Lynden-Bell, F., 1986, MNRAS, 221, 1023.

Kulkarni, S.R. and Heiles, C., 1988, in *Galactic and Extragalactic Radio Astronomy*, 2nd ed., *ibid.*, p. 95.

Lockman, F.J. and Gehman, C., 1991, *Ap. J.*, 382, 182.

McCammon, D. and Sanders, W.T., 1990, A.R.A.A. 28, 657.

McKee, C.F. and Ostriker, J.P., 1977, Ap. J. 218, 148.

Merrifield, M.R., 1992, A.J. 103, 1552.

Nordgren, T.E., Cordes, J.M., and Terzian, Y., 1992, A.J. 104, 1465.

Norman, C.A. and Ikeuchi, S., 1989, Ap.J. 345, 372.

Oort, J.H., 1932, Bull. Astron. Inst. Neth., 6, 249.

Parker, E.N., 1966, Ap. J. 145, 811.

Spitzer, L. Jr., 1977, *Physical Processes in the Interstellar Medium,* (New York : John Wiley).

Spitzer, L. Jr., 1990, A.R.A.A. 28, 71.

Sullivan, W.T., 1984, *The Early Years of Radio Astronomy,* Cambridge : University Press.

Turner, B.E., 1988, in *Galactic and Extragalactic Radio Astronomy*, 2nd ed., *ibid.*, p. 154.

van de Hulst, H.C., Muller, C.A., and Oort, J.H., 1954, Bull. Astron. Inst. Neth. 12, 117.

Yamauchi, S., et al., 1990, Ap. J. 365, 532.

Zweibel, E.G., 1987, in *Interstellar Processes*, ed. D.J. Hollenbach and H.A. Thronson, Jr. (Dordrecht : Reidel). pp. 195-221

The Minnesota Lectures on Clusters of Galaxies and Large-Scale Structure
ASP Conference Series, Vol. 39, 1993
Roberta M Humphreys (ed.)

GALACTIC DYNAMICS

IVAN R. KING
Astronomy Department, University of California, Berkeley, CA 94720

INTRODUCTION

The subject of Galactic Dynamics is large and complicated—as is the Galaxy itself. I will try here to give a simplified view of the dynamics of our Galaxy, and to do so I will make some simplifying assumptions that are customarily made in an introduction to the problem. First, we will consider the stars not as individuals but as a continuous distribution function of positions and velocities. Their gravitational potential will be assumed also to be smooth, so that we do not need to consider the individual interactions of stars on each other—the phenomena called stellar encounters. Finally, we will consider the Galaxy to be axially symmetrical and in a steady state. This last is probably the least tenable of our assumptions, but it leads to a good starting basis from which to consider the dynamics of a more realistic, more complicated Galaxy.

Occasional exceptions will be made to these simplifications, but they will be identified.

This account, arising as it does from a single pair of lectures, will take up only the high points, attempting to concentrate on the basic ideas and quoting the mathematical results without proof, and with only the barest concession toward indicating where they come from. For more details the reader is referred to an elementary discussion (Gilmore *et al.* 1990), a basic introduction (King 1993), and an advanced treatise (Binney and Tremaine 1987).

THE COLLISIONLESS BOLTZMANN EQUATION

To examine the dynamical behavior of stars in a stellar system, we define a 6-dimensional phase space (called in statistical mechanics a μ-space) that is created by conjoining the 3-dimensional position space with the 3-dimensional velocity space. Each star is represented at any given moment by a point in this space. (But I shall casually refer to the motions of the stars themselves through the phase space.) Because of the smoothness assumptions that we are making, we can describe the distribution of stars in the phase space by a smooth density f, which is a function of the position coordinates x, y, and z and the velocity coordinates u, v, and w.

The stars are individually conserved. (Thus we are ignoring births and deaths, which actually have little effect at the scale of Galactic dynamics.) This means that their flow through the phase space must obey an equation of continuity. This equation is derived just like other equations of continuity; its form is

$$0 = \frac{\partial f}{\partial t} + u\frac{\partial f}{\partial x} + v\frac{\partial f}{\partial y} + w\frac{\partial f}{\partial z} - \frac{\partial \psi}{\partial x}\frac{\partial f}{\partial u} - \frac{\partial \psi}{\partial y}\frac{\partial f}{\partial v} - \frac{\partial \psi}{\partial z}\frac{\partial f}{\partial w}, \tag{1}$$

where $\psi(x, y, z, t)$ is the potential function, whose negative gradient yields the gravitational acceleration of a particle.

Eq. (1) is called the collisionless Boltzmann equation, since it is precisely Boltzmann's equation of kinetic theory with the term omitted that takes into account collisions between particles.

This equation has an important direct physical interpretation, which becomes clear if we compare it with the equations of motion of a star,

$$\frac{dx}{dt} = u, \quad \frac{dy}{dt} = v, \quad \frac{dz}{dt} = w, \quad \frac{du}{dt} = -\frac{\partial \psi}{\partial x}, \quad \frac{dv}{dt} = -\frac{\partial \psi}{\partial y}, \quad \frac{dw}{dt} = -\frac{\partial \psi}{\partial z}. \tag{2}$$

The terms on the right of Eq. (1) contain all possible partial derivatives of f with respect to the independent variables, and each turns out to be multiplied by a quantity that is equal to the derivative of that variable with respect to t, along the path of motion of an individual point. Thus Eq. (1) states that along the path of any star through the phase space, the total derivative of f (what is often referred to in hydrodynamics as the "Lagrangian derivative") is zero. The equation is thus equivalent to what in statistical mechanics is called Liouville's theorem: as the point representing a star moves through the phase space, the density around it remains constant. In other words, the flow of stars through the six-dimensional phase space is incompressible. (This is *not* true of the flow through the three-dimensional position space!) It is from this property that a great deal of the value of this approach derives; in effect, the stars at any given position and velocity are carrying with them, in their phase density, information about the phase density at other velocities and at other spatial points. In a certain sense, it is because of this property that we are able to use the velocity distribution of local stars to probe the characteristics of remoter parts of our Galaxy.

The collisionless Boltzmann equation is fortunately of a type whose general solution is known; it is a linear first-order partial differential equation. (It is not even, in fact, the most general form of such an equation, in which a function of the variables would appear on the left-hand side too.) The general solution of this type of equation will be given here without a proof, which can be supplied by any of the numerous standard texts on partial differential equations.

Consider the equation

$$A_1 \frac{\partial f}{\partial x_1} + A_2 \frac{\partial f}{\partial x_2} + \cdots + A_n \frac{\partial f}{\partial x_n} = 0, \tag{3}$$

where the A_i are functions of $x_1, x_2, \ldots, x_n$. To find the general solution, form the "subsidiary equations"

$$\frac{dx_1}{A_1} = \frac{dx_2}{A_2} = \cdots = \frac{dx_n}{A_n}. \tag{4}$$

These are $n-1$ independent ordinary differential equations. The solution algorithm says to find $n-1$ independent integrals of these ordinary differential equations; let them be expressed in the form

$$I_i(x_1, x_2, \ldots, x_n) = \text{const.}, \quad i = 1, 2, \ldots, n-1. \tag{5}$$

Then the general solution of Eq. (3) is

$$f(x_1, x_2, \ldots, x_n) = F(I_1, I_2, \ldots, I_{n-1}), \tag{6}$$

where F is an arbitrary function of its arguments. (This is, of course, in accord with the principle that where the general solutions of ordinary differential equations have arbitrary constants, the general solutions of partial differential equations instead have arbitrary functions.) The solution given by Eq. (6) may not at first appear to offer much information, since it contains a function that is completely arbitrary; but it is in fact a strong restriction, since it says that what might have appeared to be a function of n variables must in fact depend on only $n-1$ variables. We shall soon see, furthermore, how this restriction can become even stronger.

If one writes down the subsidiary equations for the collisionless Boltzmann equation, they can easily be seen to be equivalent to the equations of motion of an individual star. This means that the theorem about partial differential equations says that the general solution of the collisionless Boltzmann equation requires that f be expressible as a function, F, of the integrals of the equations of motion of a star. But these integrals, taken all together, define the motion of the star through phase space. Thus f remains constant as the star moves. But this is just a repetition of Liouville's theorem. We have recovered the same result via this different route; but of course it had to be so, if our whole picture is to be consistent.

There are two points particularly to be noted here. First, let it be clear (and clearly remembered) that the integrals referred to are integrals of the motion of an individual test star, and have nothing to do with collective integrals, such as the conserved energy of the whole N-body system. Second, note clearly that the conserved density that is referred to is a density in the six-dimensional phase space, not in the position space.

This much we have done in a general context. But now we apply the assumption, already mentioned, that the Galaxy is in a steady state. An immediate consequence is that because f is constant along a path, then it must have the same value, at all times, at every point on the path of a star in the phase space. Thus in a sense the stellar orbits map out the distribution function.

We will follow this idea through mathematically. The collisionless Boltzmann equation now lacks the time derivative, but is otherwise unchanged. The subsidiary equations are easily written down again, and they readily yield the familiar energy integral,

$$I_1 = \tfrac{1}{2}(u^2 + v^2 + w^2) + \psi(x, y, z) = \text{const.}, \tag{7}$$

which we shall often refer to familiarly as E. It is a well-know fact of dynamics that if the potential is independent of time, this integral always exists, whatever the form of the potential may be. (Note that this integral, in the form in which we use it, is actually energy per unit mass. For ease of reference, however, we shall call it simply energy.)

INTEGRALS AND JEANS' THEOREM

Suppose now that the energy integral were the only integral that existed—that is, the only quantity conserved along the orbit of a star. If we consider the six-dimensional phase space from the point of view of analytic geometry, then $I_1 = \text{const}$ represents a five-dimensional hypersurface. Since I_1 is conserved, the motion of the star must be confined to that surface, since it could go nowhere else without having a different value of I_1.

Granted the confinement to this energy surface, let us consider now the other characteristics of the motion of the star. If there are no other restrictions on its motion, then it can be expected eventually to visit every point on the whole energy surface. In this case the value of the phase-space density f must then be constant everywhere on that surface. On any other energy surface, however, f can have some other value. No two energy surfaces can intersect, because no point in the phase space can correspond to two different values of the energy. Thus each energy surface has its own value of the phase-space density f. The simple mathematical way of summing up these statements is

$$f(x, y, z, u, v, w) = F(I_1). \tag{8}$$

This is an amazingly strong result, and we must examine how it could possibly be true. Our general solution said that F should have as its independent variables five integrals, not just one; where have the others gone? The answer to this question depends on the distinction between *isolating* and non-isolating integrals. I_1 can properly be called "isolating," because each value of the quantity I_1 isolates a surface characterized by that particular value, and the possible values of I_1 decompose the phase space into a disjoint set of energy surfaces. Whether other integrals do this or not depends on their individual nature. In fact, when the potential has no special characteristics, the powerful restriction expressed by Eq. (8) really does hold.

There are dynamical situations, however, in which it is easily shown that another isolating integral exists. For example, elementary mechanics shows that an axisymmetric potential always has an integral that expresses the conservation of the axial component of angular momentum. Let us call it $I_2 = \text{const}$. Again, from the point of view of analytic geometry this can be considered to be the equation of a five-dimensional hypersurface, and the motion must be confined to it. But the motion is also confined to a particular I_1 surface; so it must in fact take place on the four-dimensional hypersurface that is the intersection of the I_1 surface and the I_2 surface. Again, Liouville's theorem requires that f have the same value everywhere on this surface, and the mathematical expression of this fact is

$$f(x, y, z, u, v, w) = F(I_1, I_2). \tag{9}$$

The general principle exemplified by these two cases was stated a long time ago by Jeans: In a steady state, the phase-space density distribution is a function of the isolating integrals of the motion of a test star—and of *only* the isolating integrals. Jeans' theorem is one of the cornerstones of stellar dynamics.

The crux of the problem is clearly how many isolating integrals exist. Several specific forms of the potential have additional isolating integrals. As we will see, however, the problem is not this simple. In addition to the truly isolating integrals, others appear to exist for many or most orbits, in most dynamical problems. To pursue this train of investigation, though, we shall find it most convenient to treat the collisionless Boltzmann equation in a system of coordinates that is suitable for discussions of local Galactic dynamics, which is the context in which the fascinating question of "non-classic integrals" first arose.

GALACTIC COORDINATES AND VELOCITIES

There is a coordinate system whose use is fairly standard in discussions of the local dynamics of the Milky Way. For the position coordinates we use a cylindrical polar system R, θ, and z. For the velocity coordinates we take Cartesian but non-inertial components, each in the increasing direction of the corresponding position coordinate, and we use as symbols the corresponding capital letters U, V, W. These are related to the positional derivatives by the equations

$$U = \frac{dR}{dt}, \quad V = R\frac{d\theta}{dt}, \quad W = \frac{dz}{dt}. \tag{10}$$

Their non-inertial character arises, of course, from the fact that the directions of U and V change as a star moves along its orbit.

Converting the collisionless Boltzmann equation into these coordinates is straightforward but tedious, and we will simply give the result for an axisymmetric, steady-state stellar system, the condition that we will assume for the Milky Way. Its form is

$$0 = U\frac{\partial f}{\partial R} + W\frac{\partial f}{\partial z} - \left(\frac{\partial \psi}{\partial R} - \frac{V^2}{R}\right)\frac{\partial f}{\partial U} - \frac{UV}{R}\frac{\partial f}{\partial V} - \frac{\partial \psi}{\partial z}\frac{\partial f}{W}. \tag{11}$$

Eq. (11) is the equation with which we will do most of our Galactic dynamics.

For its general solution, we write down the subsidiary equations,

$$\frac{dR}{U} = \frac{dz}{W} = \frac{dU}{(V^2/R - \partial\psi/\partial R)} = \frac{dV}{-UV/R} = \frac{dW}{-\partial\psi/\partial z}. \tag{12}$$

These are again equivalent to the equations of motion of a star. We could integrate combinations of these equations to derive the integrals, but we already know them to be the energy integral

$$E = \tfrac{1}{2}(U^2 + V^2 + W^2) + \psi(R, z), \tag{13}$$

and the angular-momentum integral

$$h = RV. \tag{14}$$

These are the only two isolating integrals that can be found for all orbits in a general steady-state axisymmetric potential.

THE THIRD INTEGRAL

Application of Jeans' theorem now says that

$$f(R, z, U, V, W) = F(E, h). \tag{15}$$

But this immediately makes a strong prediction: whereas the dependence of f on V can be quite arbitrary, because V appears independently in h, the roles of U and W must go identically in lock-step, since they enter only in the combination $(U^2 + W^2)$, in E. More specifically, this requires that the velocity dispersions of U and W be equal—which observationally they are not, by a factor of about 2.

Clearly something is wrong here. A first thought might be that the Galaxy is not yet in a steady equilibrium, but after some fifty revolutions this seems quite unlikely. The more likely avenue to pursue is that there is a third integral of the motion of a star—or, at the very least, a quantity that is conserved for all practical purposes for a time of the order of the present age of the Milky Way. It is just this view that Oort and Lindblad took in 1926–28, when they were laying the foundations of the theory of Galactic dynamics. For stars that do not stray far from the Galactic plane, the z-force, which depends on local mass density, is effectively decoupled from the R-force, which comes from the matter all the way from here to the Galactic center. But if the z-force is independent, then the potential has two independent parts and can be written

$$\psi(R, z) = \psi_1(R) + \psi_2(z). \tag{16}$$

This separability gives rise to an additional integral for the z-motions:

$$I_3 = \tfrac{1}{2}W^2 + \psi_2(z), \tag{17}$$

a decoupled z-energy. With I_3 as a third argument of F, the dependence of the distribution function on W can now be quite different from its dependence on U.

Since most stars in the solar neighborhood remain in the low-z realm where Eq. (17) is a good approximation, this rationalization was complacently accepted for several decades. The structure began to crumble only when the increasing power of computers made it possible to calculate orbits, in a reasonable approximation to the Galactic potential, for stars whose motions above and below the Galactic plane had amplitudes of a kiloparsec or more (Contopoulos 1958). These orbits clearly conserved a quantity that resembled the z-energy, but that had a more complicated behavior and no obvious mathematical expression. A large set of Galactic orbits of stars was computed, and nearly all of them had a "third integral," as evidenced by the fact that they did not fill all of the parts of position space that were available to them. Figure 1 shows an example of such an orbit.

Strictly, the check should be made in phase space. This can be done by means of a diagram known as a "surface of section," created by plotting, for all crossings of the Galactic plane, the U component of velocity against the R of the crossing point. For an ergodic orbit, crossings at a given R should have all values of U that are consistent with the E and h of the orbit in question. This is not the case, however. For nearly all orbits that have been calculated, the

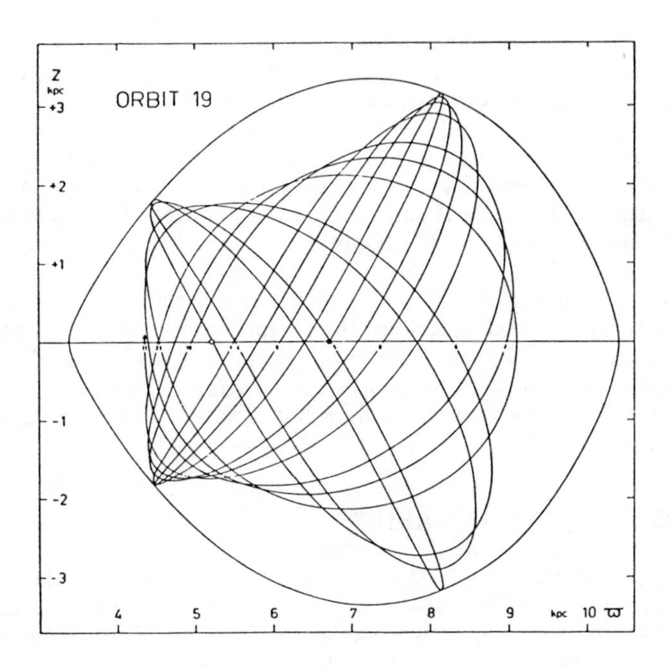

FIGURE 1 An orbit, in cross section in the (R, z) plane (Ollongren 1965; note that R is here labeled ϖ, and that the θ coordinate is ignored). The enveloping oval shows the region that is accessible to a star with the particular values of E and h that pertain to this orbit.

points in the surface of section do not fill all allowable U's at each R; instead they lie on a single "invariant curve" in the surface of section—showing that the orbit is non-ergodic, or as the terminology goes in orbit classification, "regular."

This third integral remained puzzling, however, because it lacked any identifiable properties of a real physical integral. Could it be that it was some kind of artifact of the way in which the potential and its gradient had been calculated? To check on such a possibility, Hénon and Heiles (1964) calculated orbits in a two-dimensional potential that was far too simple to have any hidden pitfalls. Would it show an additional integral, besides the obvious energy integral? When they calculated orbits at a low energy, every orbit turned out to have a second integral. This was strange enough, but then they encountered another surprise. At higher values of the energy some of their orbits jumped all over the surface of section in an ergodic way; instead of being regular, they were "stochastic." Furthermore, the fraction of stochastic orbits increased rapidly with increasing energy, until at high energy all orbits were stochastic.

At this point the classical idea of an integral was no longer applicable in any form. Here was a quantity that was conserved for some starting conditions but not for others. Although it shares some of the properties of a true integral, it can be so only in some sort of asymptotic sense. Such an integral is called a non-classical integral.

The investigation of Hénon and Heiles was a turning point in the understanding of dynamical systems. Since then, nearly every dynamical system that has been investigated has turned out to have a non-classical integral. In another direction, the investigation of the properties of their stochastic orbits was one of the starting points of the new branch of mathematics that has become known as chaos theory.

LOCAL GALACTIC DYNAMICS

With these preliminaries, we return to the main stream of Galactic dynamics. For the study of the local dynamics of the Milky Way, several different approaches are possible. Here we shall examine three of them, in some cases reaching the same result over again by different methods. Each of the approaches has its advantages, and its weaknesses, and each offers some unique insights into the dynamics of the Milky Way.

The Oort Substitution

One possibility is to assume a particular mathematical form for the velocity distribution and to see how its parameters vary from point to point. This was the approach used by Oort when he developed much of the basic picture of Galactic dynamics, in the investigation of the characteristics of a rotating Galaxy that he carried out in the middle and late 1920's.

Oort's approach was to assume that the velocity distribution had a trivariate Gaussian form:

$$f = f_0 \exp(-Q), \tag{18}$$

where

$$f_0 = h\,k\,l\,\nu/\pi^{\frac{3}{2}}, \tag{19}$$

$$Q = h^2U^2 + k^2(V - V_0)^2 + l^2W^2 + mU(V - V_0) + nUW + p(V - V_0)W, \quad (20)$$

and h, k, l, m, n, p, V_0, f_0 are functions of R and z. Thus we have assumed that the velocity distribution is a trivariate Gaussian everywhere, but that its size, shape, orientation, and mean rotational velocity can vary from place to place. Note that the second equation relates f_0 to a normalized Gaussian distribution, so that ν is the space density of stars (which we have elsewhere denoted by N).

In the solar neighborhood this is a fair approximation, and it is reasonable to assume that the velocity distribution takes a similar form at other locations in the Milky Way. The constants h, k, l, ..., however, will depend on position. Oort's approach was to substitute the form given by Equations (18)–(20) into the steady-state, axially symmetric collisionless Boltzmann equation,

$$0 = U\frac{\partial f}{\partial R} + W\frac{\partial f}{\partial z} + \left(\frac{V^2}{R} + K_R\right)\frac{\partial f}{\partial U} - \frac{UV}{R}\frac{\partial f}{\partial V} + K_z\frac{\partial f}{\partial W}, \quad (21)$$

and see what the consequences are for the spatial behavior of the constants.

There are two small points to note here. First, we are using Oort's notation for acceleration, $\boldsymbol{K}$, instead of the negative gradient of the potential. Second, following Oort we have tacitly omitted from Q the quantities U_0 and W_0; they could have been included and, at the cost of some small additional complication, shown to be equal to zero.

When f is substituted into Eq. (21), the result is a large number of terms, involving all possible powers of U, V, and W up to third-power terms. Since the resulting equation must hold for all values of the velocity components, not just for three values of U for each pair of values of V and W, etc., the supposed equation must be an identity, so that when the terms are grouped in powers of U, V, and W, the coefficient of each power must be equal to zero. Many of the resulting 21 equations involve spatial derivatives of h, k, l, ... After a complicated sequence of integrations, the results are

$$h^2 = c_1 + \tfrac{1}{2}z^2, \quad (22)$$

$$k^2 = c_1 + c_2R^2 + \tfrac{1}{2}c_5z^2, \quad (23)$$

$$l^2 = c_4 + \tfrac{1}{2}c_5R^2, \quad (24)$$

$$n = -c_5Rz, \quad (25)$$

$$m = p = 0, \quad (26)$$

$$V_0 = \frac{c_3R}{c_1 + c_2R^2 + \tfrac{1}{2}c_5z^2}. \quad (27)$$

The first five equations give strong information about the velocity ellipsoid. First, the fact that m and p are zero shows that one axis of the velocity ellipsoid always points toward the axis of rotation of the Galaxy and another axis in the direction of rotation. Moreover, at $z = 0$ the first axis points along the center–anticenter line and the third axis perpendicular to the plane; but away from the plane these axes may tilt.

It is well known, however, that many types of stars have velocity ellipsoids whose long axis does not point toward the axis of rotation. The present theory does not allow such a "vertex deviation," which is presumably due to departures from a steady state, from axial symmetry, or both.

The first three equations also tell how the velocity dispersions vary with position. Since Equations (18) and (20) define a Gaussian distribution, it follows that $h^2 = 1/2\sigma_R^2$, etc. Thus Eq. (22) tells us that the U dispersion does not depend on R (or "is isothermal in R"), although it will change with z; similarly, Eq. (24) says that σ_z is isothermal in z but should change with R.

Here we have a contradiction with observational fact. It has long been known that $\langle Z^2 \rangle$ increases with increasing distance from the Galactic plane, and Lewis and Freeman (1989) have recently shown that $\langle U^2 \rangle$ decreases with increasing R. In studies of z-motions it has therefore been customary to employ a sum of Gaussians of the type discussed here (Oort 1932, 1960). Radial studies have not yet used this technique, but it is clear that any application of Gaussian velocity distributions would have to do so. The isothermality properties exist in any given component; but the relative proportions of the components vary spatially, and so, accordingly, do the velocity dispersions.

The behavior of σ_θ is different; it is connected with Galactic rotation. We investigate it by using Eq. (27) at $z = 0$, expressing the Oort A and B constants,

$$A = -\tfrac{1}{2} R \frac{d}{dR}\left(\frac{V_0}{R}\right), \qquad (28)$$

$$B = A - V_0/R, \qquad (29)$$

in terms of the c's and R and z, and noting that h and k are inversely related to the velocity dispersions. The result is

$$\frac{\sigma_\theta^2}{\sigma_R^2} = \frac{-B}{A - B}. \qquad (30)$$

Thus the ratio of the U and V axes of the velocity ellipsoid is directly connected with the law of differential rotation of the Galaxy. Eq. (30), which will follow also from our other treatments of galactic dynamics, is often referred to as "the star-streaming equation," using the name that Kapteyn gave to this first velocity anisotropy ever discovered.

The manipulations that resulted in Equations (22)–(27) used nearly all of the powers of the velocities in the expression that resulted from the Oort substitution. Two of the powers give a different result, however. They contain the spatial gradients of f_0, which corresponds closely to the star density; from these we can derive information about density gradients. For the radial density gradient, the equation that results is

$$V_c^2 - V_0^2 = \frac{1}{2h^2}\left(-\frac{\partial \ln \nu}{\partial \ln R} - 1 + \frac{h^2}{k^2} - \frac{1}{2}\frac{c_5 R^2}{l^2}\right). \qquad (31)$$

Here we have replaced K_R by its equivalent $-V_c^2/R$, where V_c is the circular velocity at the Sun's distance from the Galactic center.

The left side of Eq. (31) expresses a lag of the mean velocity of a group of stars behind the local circular velocity. Because of the association of this lag with the general tendency for the distribution of V velocities to be different in the forward and backward directions, we shall refer to Eq. (31) as "the asymmetric-drift equation." It has important applications, which we will discuss after rederiving it in a sounder way by the velocity-moment technique. That equation will be directly applicable, whereas Eq. (31) requires, as already indicated, a more complicated superposition of solutions.

Another of our conclusions needs modification too. In Eq. (27) we have in fact derived the whole rotation curve of the Galaxy, from its center out to infinity. (This form is sometimes called a Chandrasekhar rotation curve, from its derivation by that astronomer as part of a more general discussion.) Such a form is clearly too restrictive, but the contradiction is easily understood by examining the extent to which our original assumption about the velocity distribution can be taken seriously. What is literally true is that if the *whole* velocity distribution were described by Equations (18) and (20), then the rotation curve would have to take exactly the form of Eq. (27). The fallacy is that, although Equations (18)–(20) represent the actual velocity distribution reasonably well for the lower velocities, the higher-velocity stars show all sorts of asymmetries and non-Gaussian characteristics. Here, then, is just where we can use Liouville's theorem to understand the paradox of the too strongly determined rotation curve. The stars farthest from the mean velocity satisfy the assumptions the least well. But they are the stars that go farthest from the solar neighborhood; so it is natural that our rotation curve is much less credible far from the solar neighborhood. In fact, except for a general crude shape, we can use it reliably only to relate the Oort constants to the shape of the velocity ellipsoid, as we have already done.

Finally, the assumption of a quadratic velocity distribution has one further important consequence. One manipulation of the equations that result from the Oort substitution leads to a partial differential equation that the potential function ψ must satisfy. The equation looks formidable when expressed in our present Galactic coordinates, but in confocal coordinates (in which the coordinate lines are ellipses and hyperbolas that share a common pair of foci on the axis of rotation of the Galaxy) the equation takes such a simple form that its general solution is easily found. A potential of this type is a special case of the class known as Stäckel potentials, which have a number of useful properties in stellar dynamics. In the Galactic case, approximating the actual potential of the Milky Way by a Stäckel potential allows us to find how the direction of the long axis of the velocity ellipsoid tilts as we move out of the Galactic plane.

Moment Equations

We now introduce another of the ways of manipulating the collisionless Boltzmann equation. In this approach we integrate over the velocities. Although the full truth about a stellar system is indeed contained in its full phase-space distribution function, it is rare that we have such detailed information available, and in practice the properties for which we are searching often relate only to spatial variations of density, of mean velocity, or of velocity dispersions. In order to isolate information about the spatial properties of the system, it thus makes good sense to integrate the collisionless Boltzmann equation over the velocities, term by term. The resulting terms all contain moments of the velocity distribution, usually including the zero-th moment, which is of course the spatial density. These moment equations have often been referred to under the heading of "equations of stellar hydrodynamics," but we shall avoid that term here, because it belittles the versatility of moment equations. This versatility arises from the fact that we are free to multiply the collisionless Boltzmann equation through by any powers of the velocities before integrating, and each choice of powers leads to a different result.

In the derivation of moment equations the integrations that need to be performed can be carried out by following a simple set of rules. We shall not state them here; we merely we note the general definition of a moment:

$$\langle U^j V^k W^l \rangle = \frac{\iiint_\infty U^j V^k W^l \, f \, dU \, dV \, dW}{\iiint_\infty f \, dU \, dV \, dW}, \tag{32}$$

and quote some of the results of moment integrations. The moment equations that are most useful are those that are found by integrating after multiplication by U and W, respectively. These equations are

$$0 = \frac{\partial(N\langle U^2\rangle)}{\partial R} + \frac{\partial(N\langle UW\rangle)}{\partial z} + \frac{N}{R}\Big(\langle U^2\rangle - \langle V^2\rangle\Big) + N\frac{\partial\psi}{\partial R}, \tag{33}$$

$$0 = \frac{\partial(N\langle W^2\rangle)}{\partial z} + \frac{\partial(N\langle UW\rangle)}{\partial R} + \frac{N}{R}\langle UW\rangle + N\frac{\partial\psi}{\partial z}. \tag{34}$$

These are often referred to as the stellar hydrodynamic equations, or sometimes as the Jeans equations. These equations almost separate the motions parallel to the Galactic plane from the perpendicular motions; only the $\langle UW\rangle$ terms couple the two equations. In the early development of Galactic dynamics it was customary to neglect the coupling terms and treat the parallel and perpendicular motions as completely independent; here, however, we will note that the $\langle UW\rangle$ terms in the $\langle U^2\rangle$ equation can be treated adequately by applying the theory of Stäckel potentials, whereas the corresponding terms in the $\langle W^2\rangle$ equation will raise fundamental difficulties when we take up the study of motions perpendicular to the Galactic plane.

We now treat the "U-equation," (33), in order to derive the velocity-moment form of the asymmetric-drift equation. To do this, we first make the substitutions

$$\frac{\partial\psi}{\partial R} = \frac{V_c^2}{R}, \tag{35}$$

$$\langle V^2\rangle = V_0^2 + \langle V'^2\rangle, \tag{36}$$

where V_c is the circular velocity, V_0 is the mean velocity, and V' is the residual velocity of a star with respect to this mean. [Note that Eq. (36) is a general property of all distribution functions.] After a little manipulation the result becomes

$$V_c^2 - V_0^2 = \langle U^2\rangle \left[-\frac{\partial \ln(N\langle U^2\rangle)}{\partial \ln R} - \left(1 - \frac{\langle V'^2\rangle}{\langle U^2\rangle}\right) - \frac{R}{N\langle U^2\rangle}\frac{\partial(N\langle UW\rangle)}{\partial z}\right]. \tag{37}$$

This is the moment form of the asymmetric-drift equation. It corresponds closely to Eq. (31), which we derived from the Oort substitution; but there are some subtle differences. Here we have $\langle U^2 \rangle$ instead of $1/2h^2$, but the Gaussian form of the Oort substitution made them equal in that more restrictive case. The R derivative now includes $\langle U^2 \rangle$; when we made the Oort substitution the R gradient of this quantity was found to be zero, but our present result applies to more general velocity distributions. The middle term is the same as before, on account of the close correspondences between $1/2h^2$ and $\langle U^2 \rangle$, etc., that we have noted. The last term in the brackets looks different from the one that resulted from the Oort substitution, but when these correction terms are evaluated, they turn out to be almost equivalent.

To interpret Eq. (37) we begin by noting that in actual quantitative fact the dominant term on the right is the density gradient, which is of course negative, so that both sides of the equation are positive. Thus the mean velocity of a stellar type, V_0, is less than the circular velocity V_c. For disk stars the two quantities do not differ by much, however; so we can write

$$V_c^2 - V_0^2 = (V_c + V_0)(V_c - V_0)$$

and approximate this by $2V_c(V_c - V_0)$. It then follows from Eq. (37) that the mean velocity of a group of stars lags behind the circular velocity by an amount that is proportional to its σ_R^2. This property of stellar motions is sometimes called the Strömberg quadratic relation, after the astronomer who discovered it empirically in the 1920's, before its reason was understood.

The asymmetric-drift equation is quite valuable, beyond merely showing that high-velocity-dispersion stars lag. Its quadratic character allows another important result to be extracted from it: we can locate the circular velocity, whose relation to the Sun's velocity is by no means obvious. Since the various species of stars that go to make up the Galactic disk have generally similar density gradients (as evidenced, indeed, by the very existence of the Strömberg quadratic relation), we should be able to plot the observed mean velocity against σ_R^2 for different types of stars and get a straight line. All these mean values are necessarily referred to the Sun, which is the only available reference point. The extrapolation of the line to zero velocity dispersion should then give the circular velocity, referred to the Sun.

Even more important, perhaps, is the fact that the asymmetric-drift equation allows us to find density gradients from velocities, without directly measuring densities at remote points. On the right side of Eq. (37) the other terms can be evaluated. We have already noted that observations exist from which we can evaluate $\partial \ln\langle U^2 \rangle / \partial \ln R$. The second term is of course known from local observation, and we have also noted that the theory of Stäckel potentials can be used to approximate the final term. Furthermore, these other terms turn out to be considerably smaller in magnitude than the gradient term, so that they need not be evaluated with high precision. Thus the logarithmic density gradient $\partial \ln N / \partial \ln R$ can be found from other observed quantities. This is a direct method of evaluating the scale length of the Galactic disk—purely from the characteristics of the local velocity distribution (plus a correction for the gradient of $\langle U^2 \rangle$).

Finally, we note in passing that there is a third-moment equation that reproduces the star-streaming equation (30), along with correction terms that depend on the non-Gaussian character of the velocity distribution.

Epicycle Orbits

At the same time, in the 1920's, that Oort was developing the consequences of Galactic rotation by using the collisionless Boltzmann equation, Lindblad was pursuing the same astronomical questions by the quite different technique of investigating how neighboring stars move relative to each other, as they all go around the Galactic center.

The first step is to restate the Newtonian equations of motion in polar coordinates with respect to the Galactic center. The second step is to take a reference point that moves in a circular orbit around the Galactic center, with the appropriate circular velocity. Lindblad then chooses a set of local coordinates ξ and η around the reference point and finds, to first order, the motion of a star with respect to the reference point. His first-order equations of motion for the star become

$$\frac{d^2\xi}{dt^2} - 2\omega\frac{d\eta}{dt} - 4\omega A\xi = 0, \quad (38)$$

$$\frac{d^2\eta}{dt^2} + 2\omega\frac{d\xi}{dt} = 0, \quad (39)$$

These are the equations of motion of a particle with respect to a reference point that moves in a circular orbit.

Lindblad used an approach like this to follow the relative orbits of stars in a rotating galaxy, and he was able in that way to derive many of the same results that Oort found by such a different method. We shall discuss Lindblad's orbital solution here partly to demonstrate it as an independent approach to Galactic dynamics, partly because it demonstrates a natural oscillation frequency of the Galactic disk, and partly because of the intrinsic value of the resulting orbits in following the motions of young stars.

The equations (38) are easily solved to give the result

$$\xi = c_1 + c\cos\kappa(t - t_0), \quad (40)$$

$$\eta = -2Ac_1(t - t_1) - \beta c\sin\kappa(t - t_0), \quad (41)$$

where

$$\kappa = \sqrt{4\omega(\omega - A)} \quad (42)$$

$$= \sqrt{-4B(A - B)} \quad (43)$$

and

$$\beta \equiv \frac{2\omega}{\kappa} \quad (44)$$

$$= \sqrt{\frac{\omega}{\omega - A}} \quad (45)$$

$$= \sqrt{\frac{A - B}{-B}}. \quad (46)$$

The two second-order equations required four integrations, yielding the constants of integration c, c_1, t_0, and t_1. The "epicycle frequency" κ and the related axial ratio β are, on the other hand, properties of the Milky Way.

The equations of the epicycle can be further simplified by moving the origin from $\xi = 0$ to $\xi = c_1$ and requiring that the new origin move with the circular velocity that is appropriate for that distance from the Galactic center. The leading constant term in the ξ equation then disappears, and the fact that the differential rotation of the Milky Way has a shear rate $-2A\Delta\xi$ causes the first term in the expression for η likewise to disappear. The solutions then take the simple form

$$\xi = c \cos \kappa(t - t_0), \tag{47}$$
$$\eta = -\beta c \sin \kappa(t - t_0), \tag{48}$$

The orbit that is parameterized by Equations (47) and (48) is a retrograde ellipse with semi-axes c and βc, and its center moves with circular velocity. Common to all epicycles are the frequency κ and the related value of the axial ratio of the epicycles, β. With the values of the Galactic rotation constants that are most commonly used at the present time, $A = 14$ and $B = -12$ km sec^{-1} kpc^{-1}, the epicycle period $2\pi/\kappa$ is 1.74×10^8 years, and the axial ratio of the velocity ellipsoid (or of the epicycle) is 1.47.

Thus our general result is that every star follows an elliptical epicycle orbit around a center that moves in a circular orbit. The amplitude of the epicycle is a characteristic of the star, but its axial ratio is determined only by the rotation curve of the Galaxy. The amplitude of the epicycle measures, from another point of view, how much the actual orbit of a star differs from a circular orbit. And from a global point of view the epicycle frequency κ is the characteristic vibrational frequency of local stellar motions, in directions parallel to the Galactic plane. (In this approximation, of course, the z-motions are decoupled and have a different oscillation frequency.)

The equations of epicycle orbits again reproduce the star-streaming equation (30), when the statistics are taken of all the orbits that pass through a given point.

One of the unique advantages of the epicycle approach, however, is that it gives us an easy way to convert a star's local velocity components into the characteristics of its Galactic orbit. The derivation of the transformation is straightforward, as follows: At $t = 0$, let $\xi = \eta = 0$, while $u = u_0$ and $v = v_0$; substitute these into the epicycle equations and find expressions for ξ and η at any later time. The resulting form of the equations, derived by Blaauw (1952), is

$$\xi = \frac{1}{\kappa}\left[u_0 \sin \kappa t + v_0 \beta(1 - \cos \kappa t)\right] \tag{49}$$
$$\eta = \frac{1}{2B}\left[u_0(1 - \cos \kappa t) + v_0(2At - \beta \sin \kappa t)\right]. \tag{50}$$

Blaauw's equations conveniently allow us to go from any star's observed velocity components to its epicycle orbit. A very instructive application of them is to follow the dispersal of a newly born stellar association, in which the stars have started in life together but are gravitationally unbound and will drift apart.

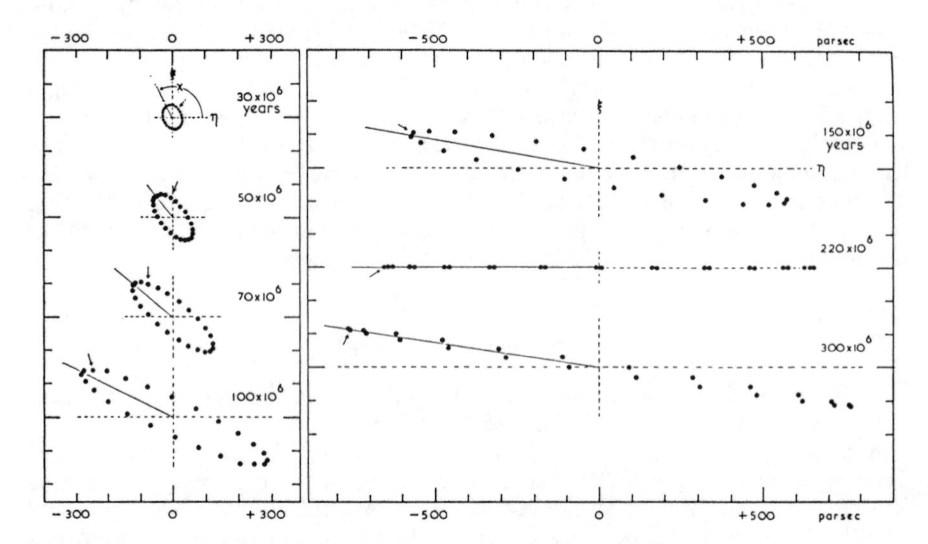

FIGURE 2 The dispersal of an association that begins expanding with relative velocities of 1 km sec^{-1}, as calculated by Blaauw (1952). Arrows mark the star that originally was moving in the direction of the $+\eta$ axis. These diagrams are calculated with the values $A = 20$ km sec^{-1} kpc^{-1}, $B = -7$ km sec^{-1} kpc^{-1}.

Blaauw derived his equations, in fact, to study just this problem. In his 1952 paper he followed the configuration of a group of stars that begin expanding from a single point, in all directions, with relative velocities of 1 km sec^{-1}. His results are reproduced in Fig. 2. The numbers would be different now, because of different notions of the values of the rotation constants, but the qualitative picture remains the same: the group skews into a continually elongating ellipse. After one epicycle period the stars have returned to a tangential line, but the secular term in the η equation, $(A/B)v_0t$, has spread them out over $(A/B)(2\pi/\kappa)$ parsecs (since Blaauw had assumed an expansion velocity of 1 km sec^{-1}). Blaauw's spread of 660 pc, which is proportional to A/B, is reduced by modern values of the rotation constants to 270 pc, but it is nevertheless impressive to see how rapidly the shear of Galactic rotation (which is what the secular term really expresses) spreads a newly formed star group along the tangential circle, for even a single kilometer per second of expansion velocity.

MOTIONS PERPENDICULAR TO THE GALACTIC PLANE

In earlier discussions we have seen that the motions in the z direction are nearly independent of those parallel to the Galactic plane. In fact, for small z-amplitudes the Oort–Lindblad third integral becomes almost exact, and the two motions approach being totally decoupled. Since this condition is nearly met for the ordinary stars of the solar neighborhood, we shall take it as a first approximation and follow through its consequences.

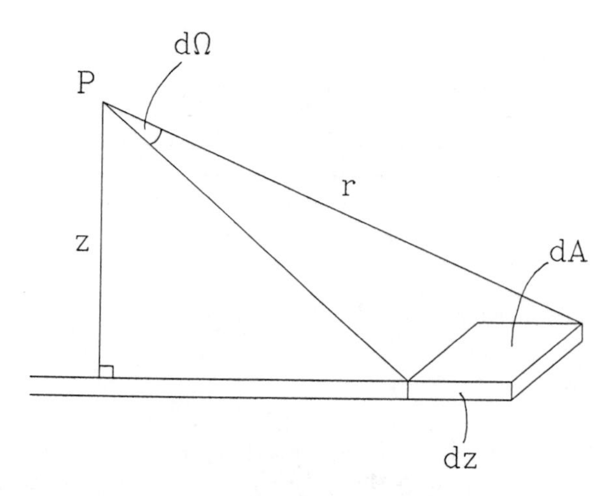

FIGURE 3 The attraction of a slab of plane-parallel stratified material.

The most important application of the analysis of z-motions is to determine the mass density in the solar neighborhood. Such an analysis was first carried out by Oort (1932), and the resulting spatial density has always been known as the *Oort limit* ("limit" because it refers to *all* of the local mass density and is thus an upper limit to the sum of the seen and the unseen).

Forces Perpendicular to a Planar Distribution

In the spirit of our local approximation, we idealize the Galactic disk as a stratification whose density depends on z but whose extent is infinite in the directions parallel to the plane. The forces due to this configuration turn out to be particularly easy to calculate, and they are a good approximation, as we shall see, to those due to the actual Galactic disk in the solar neighborhood.

As a first step, consider an infinitesimal plane slab of thickness dz but of infinite extent, with density ρ; we will calculate its attraction on a point P, at a height z above the slab. The calculation becomes very easy if we begin with that part of the force that is due to an element of area dA, at a slant distance r (Fig. 3) from P. The magnitude of the acceleration at P due to the volume element $dA\,dz$ is $G\rho\,dA\,dz/r^2$. The z-component of this is $-z/r$ times that expression (negative because the acceleration is downward). But the solid angle that dA subtends at P is easily seen to be $d\Omega = (dA/r^2)(z/r)$, so that the contribution of this element to the z-component of acceleration can be written $dK_z = -G\rho\,dz\,d\Omega$. If the plane is infinite, it subtends a solid angle 2π, and integration over this whole solid angle gives

$$dK_z = -2\pi G\rho\,dz \tag{51}$$

—a strikingly simple result, because the acceleration is independent of the distance of P from the slab.

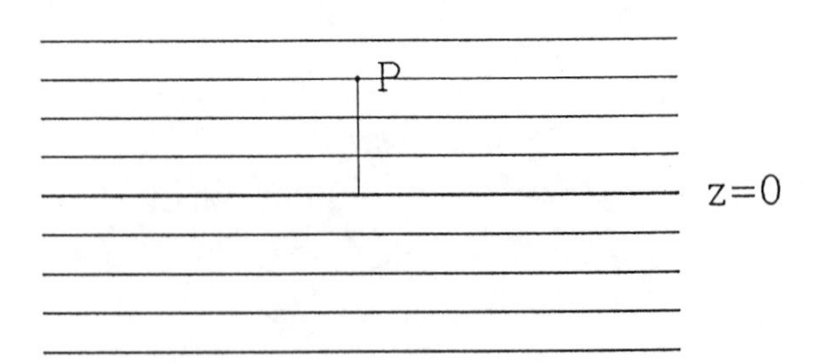

FIGURE 4 Layers of density above and below the galactic plane, schematically.

Now consider a vertical distribution $\rho(z)$, symmetrical about the central plane $z = 0$ (Fig. 4). It is clear from the figure that if z_P is the height of P then the two layers at $+z$ and $-z$ pull in the same direction when $z < z_P$ but pull oppositely and cancel each other when $z > z_P$. Thus the total acceleration is just that due to the material, on both sides of the Galactic plane, that is closer to the central plane than is P. The total acceleration at P is then

$$K_z = -4\pi G \int_0^z \rho(z)\, dz \tag{52}$$

(where we have dropped the no-longer-necessary subscript P from z, and have also used the evenness of the density function to fold the integral around zero and simplify the limits).

Application to the Local Galactic Case

Near the Galactic plane there should be a region where ρ stays nearly constant (because of its symmetry around $z = 0$); in this region $-K_z$ should rise linearly. At greater distances from the Galactic plane, however, ρ drops off; the rise of $-K_z$ slows, and finally $-K_z$ levels off effectively, at a height that is outside nearly all of the disk mass density in the vicinity of the Galactic plane. Or so it would if the Galaxy did not have a dark halo, which also makes a contribution to K_z. The latter is fortunately linear, for any reasonable behavior of dark-halo densities, within the small range of z that we need to consider; so the $-K_z$ curve continues to rise at a steady rate, even beyond the effective limits of the distribution of disk mass.

The resulting K_z curve is illustrated in Fig. 5. Its behavior is beautifully transparent: after the slope due to the dark-halo contribution is allowed for, the *slope* of $-K_z$ near $z = 0$ is $4\pi G\rho_0$, which gives the local *volume* density in the plane, while the *height* difference where it becomes parallel to the dark-halo-alone curve is $2\pi G\sigma_0$, where σ_0 is the local *surface* density of disk material, *i.e.*, the total density integrated along a line in z.

In principle the above interpretation would suffer from lack of a direct knowledge of the local density of dark-halo material, which is inferred only indirectly from the rotation curve of the Galaxy and a modeling of the density of the

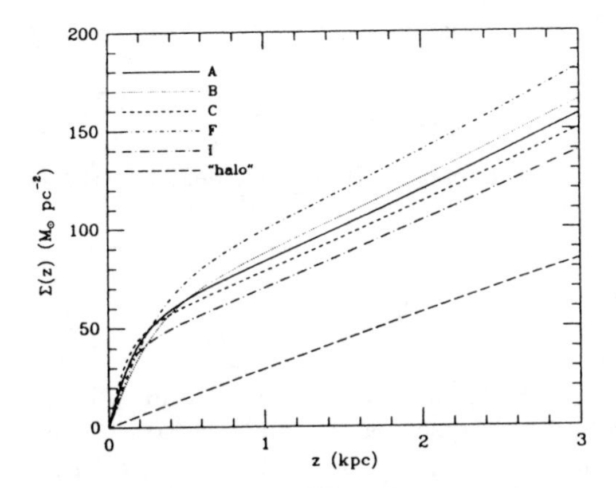

FIGURE 5 K_z curves for a variety of Galactic models, with the underlying contribution of the dark halo also shown (Statler 1989a). (Note that the slope of K_z, above the dark-halo line, is $\propto \rho_0$, while its limiting height is $\propto \sigma_0$.)

luminous material. But in practice it looks as if a good observational determination of K_z could allow the dark-halo contribution to be extracted independently. The thick disk is contributing only minimally to K_z at heights above 2 kpc, and its density is probably well enough known to allow for its contribution, thus isolating the slope of $-K_z$ that is due to the dark halo—and therefore the local density of the latter.

The above discussion depends, of course, on the approximation of constant planar densities of infinite extent. How good is this, for the range of z that we need to consider? For the solar neighborhood the main disk mass, which is due to the thin disk, has a scale height of a few hundred parsecs, and beyond a kiloparsec there is only the small contribution due to the thick disk. The approximation described above can probably be carried out to about 2 kiloparsecs. A point 2 kiloparsecs above the plane sees a density distribution that indeed has a radial drop away from the Galactic center, but for the z acceleration the higher density closer to the center is compensated by the lower density away from the center. It is only the second radial derivative of the density distribution that is damaging to our approximation. Most of the solid angle of 2π that went into our approximation is subtended by material within a few kiloparsecs of the Sun, so the approximation is not bad at all.

But in any case, let it be clear that the above discussion is presented only in a pedagogic spirit, to clarify the principles that govern the shape of the $K_z(z)$ curve. A serious calculation can abandon the planar approximation and model the densities in a more realistic way.

The reader will have noticed the hypothetical tone of this discussion of the exploitation of an observed K_z curve. The reason, as we shall see later in the chapter, is that these ideas are just beginning to be understood properly, and have not yet been applied adequately to observational data.

The Harmonic Approximation to W Motions

In the small-z region where K_z is linear, the z equation of motion of a star is

$$\frac{d^2 z}{dt^2} = -4\pi G \rho z, \tag{53}$$

which is just simple harmonic motion. It is convenient to write the solution as

$$z = z_{\max} \sin k(t - t_0) \tag{54}$$

$$W \equiv \frac{dz}{dt} = k z_{\max} \cos k(t - t_0), \tag{55}$$

where the frequency is $k = \sqrt{4\pi G \rho_0}$. (The sine form for z is chosen simply to set the zero-point of time at the passage through the Galactic plane.)

The period for the z-oscillation is $T_z = 2\pi/k = \sqrt{\pi/G\rho_0}$; with a typical value for ρ_0 of 0.15 $M_\odot/\mathrm{pc}^3$, this is 68 million years—only about a third of the rotation period or the epicycle period.

Comparison of the position and velocity equations shows that

$$z_{\max} = \frac{1}{k} W_0, \tag{56}$$

where W_0 is the value of W when the star goes through the Galactic plane. With the value just quoted for ρ_0, $z_{\max}$ in parsecs is 11.1 times W_0 in km/sec. The Sun, for instance, is very close to the Galactic plane at present and has $W_0 = +6$ km/sec; so it will reach a $z_{\max}$ of 67 pc in 17 million years.

The harmonic approximation works, of course, only for stars with small z amplitudes. Farther from the plane the force drops below this linear value; periods lengthen, and the z amplitudes become larger than our proportional equation would predict. For example, a star with $W_0 = 50$ km/sec will reach a $z_{\max}$ of about 1100 pc in 50 million years. (For a tabulation, see Oort [1965], p. 473.)

Poisson's Equation and a More Exact Density Expression

In general, Poisson's equation states that the mass density is given by

$$\frac{\partial^2 \psi}{\partial R^2} + \frac{1}{R}\frac{\partial \psi}{\partial R} + \frac{\partial^2 \psi}{\partial z^2} = 4\pi G \rho, \tag{57}$$

and not just by the $-\partial K_z/\partial z$ that we have used, which is of course equal only to $\partial^2\psi/\partial z^2$. The remaining terms in Poisson's equation turn out to be almost negligible, however. From $\partial\psi/\partial R = V^2/R$ and the expressions for the Oort constants, we find that

$$\frac{\partial^2 \psi}{\partial R^2} + \frac{1}{R}\frac{\partial \psi}{\partial R} = -2(A^2 - B^2). \tag{58}$$

Even with the old "1964 standard I.A.U. values" of A and B, this is only a few per cent of $\partial^2\psi/\partial z^2$, while with a more modern flat rotation curve $A = B$ and the correction vanishes.

Observational Determination of ρ_0

In principle the determination of the local mass density rests on the simplest of physical arguments: The density follows from the gravitational field, which can be determined by comparing the W velocities of stars with the heights that they reach above the Galactic plane. In practice, however, this is far from trivial; we can hardly throw a star upward and wait 17 million years to see how high it goes before it stops and begins to fall back. Instead, we get just a bit more sophisticated and compare, statistically, the velocity dispersion of the stars with the scale height that they reach. This is easily illustrated by making the approximation $\langle W^2 \rangle = \text{const.}$, which can be justified in two ways. First, it is not far from the truth and is pedagogically useful. (Remember, the Oort substitution actually gave that result, for any single Gaussian component.) Second, a wide range of non-Gaussian distributions can be represented by a superposition of Gaussians. (Even though Gaussians are not a mathematically complete set, the actual stellar velocity distribution can be represented rather well by such a sum.)

In this preliminary treatment we shall also omit UW terms; the nature of the complications that they introduce will be taken up later.

The hydrodynamic equation thus takes the simple form

$$\frac{\partial(N\langle W^2\rangle)}{\partial z} = -N\frac{\partial\psi}{\partial z}, \tag{59}$$

which integrates (along constant R) to give

$$N = N_0 \exp\left[\frac{-(\psi-\psi_0)}{\langle W^2\rangle}\right], \tag{60}$$

where subscript zero refers to values in the Galactic plane. Thus the velocity dispersion and the density distribution give explicitly $\Delta\psi$, whose gradient gives in turn K_z.

It is important to note that in this discussion—and throughout the problem of deducing forces from W-motions—the stars that we observe are used as test particles, whose distribution, compared with their motion, traces the gravitational field. More specifically, two points should be emphasized: (1) The distributions of density and velocity must be observed for stars of the *same type.* The two sets of data need not (and in general do not) refer to the same individual stars, but care must be taken that the velocity sample and the density sample do not differ in spectral type, abundances, or in any other physical way. (2) The tracer stars need not be in any way responsible for the gravitational field. They tell us the mass density that is present—important: the *total* mass density—but they do not tell us what kind of objects make up that mass.

In his original work (Oort 1932), as well in later discussions (Oort 1960, 1965), Oort used a simple method of approximating the actual velocity distribution by a superposition of several Gaussians. He chose the $\langle W^2\rangle$'s of the component distributions to correspond to actual astronomical types—populations of high- and low-velocity stars, and adjusted the proportions in such a way as to fit the observed velocity dispersions.

The reason that the superposition method works, of course, is that the hydrodynamic equation is linear, so that separate solutions of it can be added. As we saw above, each individual component follows the density law

$$\ln \frac{N}{N_0} = \frac{-(\psi - \psi_0)}{\langle W^2 \rangle}. \tag{61}$$

The exponential dropoff of density causes the contribution of the lower-velocity components of the mixture to be sharply reduced as z increases. The high-velocity tail of the velocity distribution gets larger, and $\langle W^2 \rangle$ increases.

In this fitting, which is done at $z = 0$, the proportion of stars of the highest-velocity-dispersion component plays a crucial role, because it is the high-velocity tail of the distribution that brings us information about conditions far from the Galactic plane (in the sense in which Liouville's theorem has been referred to several times already as a carrier of information between all regions that are visited by the same stellar orbit). In adjusting the high-velocity stars, the information of greatest leverage comes from the radial velocities of stars that are far from the Galactic plane, since at large z the exponential dependence of density on $\psi - \psi_0$ has so drastically reduced the contribution from the low-velocity components that the higher-velocity components stand out much more clearly.

In any case, for any assumed superposition of populations with different velocity dispersions it is perfectly straightforward to calculate the proportions, and thus the density and $\langle W^2 \rangle$, at any z. Oort's procedure, then, was to choose proportions of his populations in the Galactic plane such as to give the observed $\langle W^2 \rangle$, and then adjust those proportions until they give simultaneously the observed correspondence between N and z and the observed increase of $\langle W^2 \rangle$ with z.

Observational Results

The derivation of K_z, with the corresponding ρ_0 and σ_0, is straightforward in principle; but good results have never been obtained. Typically the derived $-K_z$ curve has turned down so sharply, several hundred parsecs above the plane, that it implies the absurdity of negative densities. As a result many studies have had to content themselves with a determination of ρ_0, perhaps along with a somewhat weaker value of σ_0. This was Oort's (1960) approach. He assumed the *form* of the density distribution $\rho(z)$ (and thus, implicitly, of K_z) and used the data to determine the scale factor ρ_0 which led to the best fit. Even so, he had to make some judgment-based adjustment of the observational data in order to reach any satisfactory agreement at all with the observed densities and velocities. His often-quoted result was $\rho_0 = 0.15\ M_\odot/\text{pc}^3$.

The trouble in determining K_z very probably arises from the choice of stellar types to use for this problem. In older studies the driving requirement was that the stars be luminous enough to be spectrally classified well enough, down to a magnitude that allowed their density distribution to be determined at distances up to about a kiloparsec from the Galactic plane. At the same time the stars had to be presumed to be old enough to have mixed their velocities and densities sufficiently for this time-independent theory to apply. The only stellar types that were considered to satisfy these criteria were main-sequence A stars and K giants.

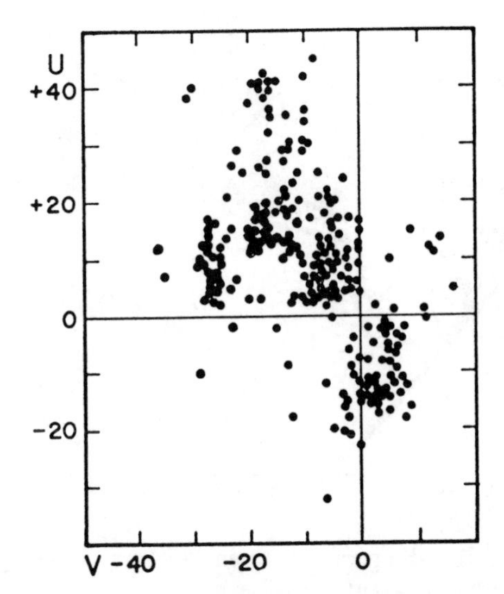

FIGURE 6 The velocity distribution of main-sequence A stars brighter than $V = 5.5$ (Eggen 1965).

Both, however, have fatal flaws. The A stars are clumpy both in velocity (see Fig. 6) and in position (McCuskey 1965, Fig. 9); they should not be describable by a time-independent equation. For the K giants the problem is inhomogeneity of populations. Because of the shapes of post-main-sequence evolutionary tracks, they "funnel" (Sandage 1957) into a region of the HR diagram where they are spectroscopically indistinguishable but actually spread over a range of absolute magnitudes. (See Fig. 7.) What is worse, absolute magnitude is correlated with age, which correlates in turn with velocity dispersion. Because, as we have seen, the z potential gradient selects out the higher-velocity stellar types at increasing z, the K giants farther from the Galactic plane should become systematically less luminous, so that their distances—and thus their densities—are mis-estimated.

Thus it appears necessary to look farther down the main sequence than the A stars. The early F's are unacceptable for the same reason as the A's. It has been recognized for 50 years that they are not uniformly distributed in the Galactic plane, with a maximum near the Sun (see, for example, Fig. 11 of McCuskey 1965); one should therefore expect them not to be in equilibrium. The late F stars are probably marginal from this point of view too, unless handled with special care. These are in fact under study by the Copenhagen group (Nordström and Andersen 1989); they are able to distinguish ages and will assume dynamical equilibrium only for the stars that their photometry labels as being old enough.

The main-sequence G stars have a different problem. Some of the oldest will have evolved upward from the main sequence, again creating a correlation between absolute magnitude and z. These differences are small but telling, and it is not clear that they can be measured adequately by the Strömgren-photometry

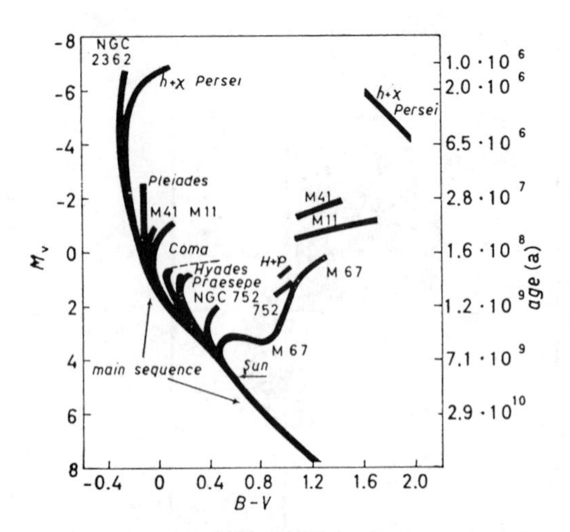

FIGURE 7 Composite HR diagrams of open clusters, showing how stars of different ages funnel into the K-giant region.

methods just referred to in connection with the F stars.

Next come the main-sequence K stars, all of which must still be unevolved. Until modern times they have received little consideration, because of their faintness; but recently Gilmore and Kuijken (1989b) have undertaken a study in the south Galactic cap that reaches down to $V \sim 17$.

Prior to this work, an interesting analysis had been carried out by Bahcall (1984) using F stars. His result for ρ_0 was 0.18 $M_\odot/\text{pc}^3$, and he optimistically claimed it to be uncertain by only 0.02. Ever since Oort's first discussion of what has come to be known as the Oort limit, results have often posed a puzzle, in that the total ρ_0 seems too large to be accounted for by known types of material. The known stars add up to about 0.06 $M_\odot/\text{pc}^3$, and the contribution of interstellar material brings this to 0.09 or 0.10. With Bahcall's ρ_0, about half of the total mass density in the immediate neighborhood of the Sun is unaccounted for. It is widely accepted that "dark matter" plays an important role in the dynamics of the Milky Way, since it is needed to account for the flat rotation curve in the outer parts of the Galaxy; but all reasonable pictures of the distribution of such a component have it so widely distributed in z that it can make only a quite small contribution to ρ_0. If there is indeed a discrepancy between the observed ρ_0 and the total local density of stars and interstellar material, then it would appear that we have more than one dark-matter problem on our hands.

The work by Kuijken and Gilmore (1989a,b,c; Kuijken 1991a), however, argues that the observations do not support any density greater than that of the observed material, which they estimate as 0.10 $M_\odot/\text{pc}^3$. A similar disagreement exists in estimates of the surface density, for which their determination is somewhat stronger. (It is worthy of note that their chief results are derived from application of a sophisticated new maximum-likelihood technique, described in the first of their three joint papers.)

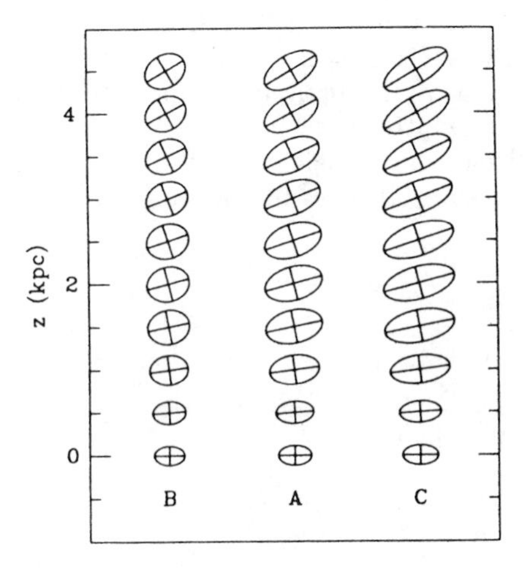

FIGURE 8 Change of the velocity ellipsoid with z for three of the models shown in Fig. 5 (Statler 1989a).

The question is far from settled. A recent conference (Philip and Lu 1989) illuminated many of its problems, however. Both Gould (1989a,b) and Statler (1989a,b) argued that the sample used by Kuijken and Gilmore, far from the Galactic plane, is statistically inadequate to provide a good determination of even the surface density, or even more so, of ρ_0. It may be that their statistical fears are somewhat exaggerated, but a far worse problem is that at these z values the UW terms in the dynamical equations become important, and their values are too uncertain. This problem is discussed with particular cogency by Statler (see 1989a, and a simplified discussion in 1989b). He shows that although the tilt of the velocity ellipsoid can be found by fitting a Stäckel potential, the spatial variation of its axis lengths is still unknown (see Fig. 8). It can be determined only by velocity observations in intermediate latitudes, which do not yet exist.

THE JEANS AND TOOMRE INSTABILITIES

Long ago Jeans (1929,[1] p. 340) took up the question of a self-gravitating gas and found that under certain conditions it could be unstable toward collapse under its own self-gravitation. His analysis considered an adiabatic gas with $p \propto \rho^{\gamma}$. We will do the equivalent thing here for a stellar system.

Conceptually we ask a simple question: will gravitation cause a configuration to collapse faster than velocity dispersion would cause it to fly apart? To answer this question we calculate a characteristic time for each process and simply assert that the faster one will win out. As might be expected, over small dimensions the velocities win; over larger scales the gravitation will predominate. Thus we

[1] In its original form the work was published many years earlier.

will find a critical length above which gravitational instability asserts itself.

Consider a spherical region of radius L and density ρ, so that its mass is $\frac{4}{3}\pi L^3\rho$. If a particle at the surface had nothing to hold it up, it would fall to the center in half the period of an infinitely narrow, elliptic orbit whose semi-major axis is $L/2$. (Everything beneath it would fall equally fast, since inside a homogeneous spherical distribution all periods are identical.) Application of Kepler's third law to such an orbit gives for the collapse time

$$T_{\text{coll}} = \sqrt{\frac{3\pi}{32G\rho}}. \tag{62}$$

On the other hand, if there were no gravitation, a star would run out to radius L in $T_{\text{esc}} = L/\langle v^2\rangle^{\frac{1}{2}}$. In the critical case these two times are equal, and the radius for which equality holds is

$$L_{\text{J}} = \sqrt{\frac{3\pi}{32}\frac{\langle v^2\rangle}{G\rho}}. \tag{63}$$

This is called the *Jeans length*; if $L > L_{\text{J}}$, then stars cannot run fast enough and collapse will occur. Or, to put it in a different way, in a stellar medium that is characterized by given values of $\langle v^2\rangle$ and ρ, all lengths $> L_{\text{J}}$ are gravitationally unstable. (Note: Jeans' analysis, for gas, had $\frac{5}{9}$ instead of $\frac{3}{32}$. The difference is not important, since we are merely finding an order of magnitude.)

The Jeans length is of great importance in discussions of the formation of galaxies, star clusters, stars, clusters of galaxies, and almost everything that might be imagined to condense in an originally more homogeneous universe. It is usually expressed not as a length, but as the *Jeans mass*, which is the size of the entities that will be chosen out by the process of gravitational instability. Our application here, however, will be to the stability of a stellar disk, and we will need to deal only with lengths.

The situation in the Milky Way is somewhat different from the problem considered by Jeans. Not only do we have a flattened rather than a round configuration, but, more importantly, there is also differential rotation. Since the latter involves velocity differences that are proportional to ΔR, it might prevent collapses from taking place if the scale is too large. If so, we have velocity dispersion to protect against collapse on small scales and differential rotation to protect against it at large scales. But what happens in between? This is what Toomre (1964) analyzed at some length. He began with an approximate, intuitive discussion and then followed up with a more rigorous analysis. We will here follow the spirit of his intuitive introduction and merely quote his more accurate result at the end.

Since the Milky Way is disk-shaped rather than spherical, we repeat the previous analysis, but with $M = \pi L^2\mu$, where μ is a surface density; but we still take the disk to attract (approximately) like a point mass. (This is not seriously in error; a sphere squashed down to infinite thinness has its equatorial attraction increased by a factor of less than 2.) Thus we have for the collapse time, as before,

$$T_{\text{coll}} = \sqrt{\frac{\pi}{8}\frac{L}{G\mu}}. \tag{64}$$

When we set this equal to $T_{\rm esc} = L/\langle v^2\rangle^{\frac{1}{2}}$, the result is

$$L_{\rm J} = \frac{\pi}{8}\frac{\langle v^2\rangle}{G\mu}; \tag{65}$$

note the different form from the spherical case. [Toomre simply asserts that what we are calling $T_{\rm coll}$ "can be estimated to be of the order of $(L/G\mu)^{\frac{1}{2}}$", so he does not have the $\pi/8$.]

Toomre's important contribution was to consider Jeans instability in the presence of differential rotation. As already indicated, the problem is now a tripartite one; the added element is that we have introduced a new fly-apart mechanism that is effective on *large* scales. What Toomre did in this first simple approach was to find the length scale beyond which differential rotation stabilizes the disk. To do this it is necessary to analyze the balance between differential rotation and self-gravitation. Differential rotation manifests itself physically through the fact that a contracting region conserves angular momentum. This spins it up and creates a centrifugal force that might be able to inhibit further contraction. The question is, as a contraction proceeds, does this centrifugal force increase faster than the gravitational force does?

Toomre answers this question by comparing the differential changes in the two forces. For the centrifugal force we can note that the formula for the Oort double sine wave in proper motions in Galactic longitude,

$$\mu_\ell = \frac{A\cos 2\ell + B}{4.74 \cdot 1000}, \tag{66}$$

indicates that the average angular velocity (with respect to a fixed system) is B. From the conservation of angular momentum the spin-up, and the consequent increase in centrifugal force, can be calculated. Similarly, the increase in gravitational force due to the decrease in size is easily calculated. The two differentials turn out to be equal if the original radius of the configuration is

$$L_{\rm rot} = \frac{2\pi G\mu}{3B^2}. \tag{67}$$

We thus find that differential rotation will stabilize a disk against gravitational collapse for length scales $L > L_{\rm rot}$.

A disk can therefore be gravitationally unstable only in the intermediate range of lengths $L_{\rm J} < L < L_{\rm rot}$—*if* this intermediate range exists at all, *i.e.*, if $L_{\rm rot} > L_{\rm J}$. This was Toomre's great recognition: that the minimum condition for stability of a disk is that $L_{\rm rot} = L_{\rm J}$. If we equate the above expressions for these two quantities, the result is

$$\langle v^2\rangle^{\frac{1}{2}} = \frac{4}{\sqrt{3}}\frac{G\mu}{B}. \tag{68}$$

If the stars of a disk have a lower velocity dispersion than this, the disk will be unstable.

Toomre's more elaborate (and complicated) analysis gave, in his notation,

$$\sigma_{u,\rm min} = 3.36\frac{G\mu}{\kappa}, \tag{69}$$

where κ is the epicycle frequency. This is called the *Toomre criterion.* (In the solar neighborhood $\sigma_{u,\text{min}}$ is ~ 30 km sec^{-1}.) If σ_u were lower, then presumably condensations would occur, and encounters between stars and these massive condensations would pump σ_u up to $\sigma_{u,\text{min}}$.

The ratio of the actual velocity dispersion to $\sigma_{u,\text{min}}$ is usually denoted by Q. It is an important quantity in the study of the dynamics of disks, and particularly in theories of spiral structure.

Note, however, that Toomre's analysis applies only to axisymmetric instabilities and that the critical length L_{T}, where $L_{\text{J}} = L_{\text{rot}}$, is many kiloparsecs. Thus violation of the Toomre criterion does not provide directly for the formation of spiral arms. Note also that disks can be subject to other instabilities, such as the formation of a central bar.

BARS AND WARPS

In the 1970's, numerical simulations of galactic disks were plagued by a tendency to form a central bar, when the intent of the researchers had been to study the behavior of a disk that was close to axial symmetry. The situation was described most strikingly by Ostriker and Peebles (1973), who suggested that the bar instability would occur in a stellar system unless the rotational kinetic energy was less than 28 percent of the total kinetic energy. They were able to prove that this is so for fluid spheroids, but for stellar systems they advanced it only as an assertion that was supported quite well by a large range of numerical simulations, done by many techniques. The existence of massive halos in galaxies was just becoming recognized, and their specific suggestion was that a disk could be "cold," *i.e.,* have a low velocity dispersion, only if a massive halo was also present, with its kinetic energy in the form of random motions.

Many further simulations have tended to confirm this conclusion. More recently, however, Athanassoula and Sellwood (1986) showed that disk stability could be achieved by increasing velocity dispersions by a reasonable amount in the central part of the disk; much less of a massive halo was then needed for stability.

Another kind of departure from symmetry is the tendency of disks to show a warp. The most distant hydrogen observed at 21 cm does not conform at all longitudes to the Galactic plane. Around $\ell = 90°$ the distant gas is north of the plane, while near 270° it is below the plane. (For details of the observations, see the review by Burton, 1988, and references therein.) Although out to the Sun's distance from the Galactic center the plane of the gas is very flat, somewhat beyond the solar circle this systematic trend begins, with a bending that increases with distance. The result is a systematic warp of the edge of the disk.

The warp has never been seen at optical wavelengths, because its material is at low latitude at a large distance, so that too much interstellar absorption intervenes. It has recently been observed in the distribution of IRAS point sources, however (Djorgovski and Sosin 1989).

As soon as the warp was discovered, more than three decades ago, it was noticed that it pointed in the direction of the Magellanic Clouds. The mass of the Clouds, however, is far too small to have such a striking effect on the potential field of the Milky Way; if the Magellanic Clouds are the cause of the warp, the interaction must operate in a more forceful, or else a more subtle way. Indeed, Hunter and Toomre (1969) were able to develop a theory in which a warp was induced by a recent close passage of the Clouds.

A warped edge is not unique to the Milky Way, however; numerous other galaxies have since been found to show warps (see, *e.g.*, Bosma 1981). Some of these galaxies have no obvious perturber. Thus is seems appropriate to seek for ways in which a galaxy can generate its own warp in a natural way. A new range of possibilities became available when it was discovered that the Milky Way has a large outer mass in its dark halo.

The bending modes of a self-gravitating disk, inside an additional potential, pose a difficult mathematical problem. Sparke and Casertano (1988) have worked out a theory of natural bending modes in a galaxy whose disk is in a plane that is tilted somewhat relative to the symmetry plane of the galaxy's somewhat flattened massive halo. They find that two types of bending are possible, depending on the core radius of the halo, thus raising the intriguing possibility that (if this theory is right) the shape of a warp can tell us otherwise unobservable facts about the dark halo. More recently, Kuijken (1991b) has extended such a theory into the non-linear domain.

THE INCREASE OF VELOCITY DISPERSION WITH TIME

Ever since velocity dispersions of stars were first measured, it has been clear that different types of stars have different velocity dispersions. The early tendency was to compare stars by spectral type; it was clear that the late-type stars had higher velocity dispersions than did the stars of early type. The crucial step in interpreting the velocities was the recognition that the G, K, and M stars actually have quite similar velocity dispersions. Thus velocities increase from the O stars up to the G stars and then level off. This break, often called Parenago's discontinuity, occurs near the place where the main-sequence lifetime equals the age of the Galactic disk.

Now all becomes clear. What the velocity dispersions correlate with is the mean age of the group of stars. The question then becomes, what dynamical mechanism causes velocity dispersions to increase with age?

Several decades ago, Spitzer and Schwarzschild (1951, 1953) suggested that the velocities of old stars might be increased over time by encounters with massive interstellar complexes. (In the modern context, we would tend to think of giant molecular clouds.)

Their basic idea becomes clear if we examine Chandrasekhar's (1960; see Eq. 2.379) formula for the time of relaxation as a result of stellar encounters:

$$T_{\mathrm{E}} = \frac{1}{16}\sqrt{\frac{3}{\pi}}\frac{v_1^3}{N_2 G^2 m_2^2 \ln[D_0 v^2/G(m_1+m_2)]}. \tag{70}$$

Subscripts 1 and 2 refer to the objects studied and the objects encountered, respectively. The v's are velocity dispersions; the unsubscripted one is a typical relative velocity in an encounter. The encounter cutoff length D_0 can be taken to be the scale height of the disk; its exact value is unimportant, because the argument of the logarithm is so large that the value of the logarithm is quite insensitive to it. N_2 is the number density of objects encountered.

For ordinary encounters between stars in the solar neighborhood relaxation is far too slow to matter; for stars encountering stars, T_E is 10^{13} to 10^{14} years. But when the objects encountered are massive, the situation changes, because of the $N_2 m_2^2$ factor in T_E; remember that "2" refers to the *other* object. If we note that $N_2 m_2^2$ can be written as $m_2 \cdot N_2 m_2 = m_2 \rho_2$ (where ρ_2 is the mean spatial mass density of objects of type 2), then it is clear that the velocity-diffusion effect due to objects with a given average density is *proportional to the mass of an individual object.* Thus interstellar material collected into clouds of size $10^5 M_\odot$ will be 10^5 times as effective in changing stellar velocities as the same amount of material would be if it were in chunks of one solar mass each. Note, however, that the presence of the v_1^3 factor makes it increasingly hard to pump up a velocity dispersion once it has begun to rise. (This is why such arguments do not apply to the stars of the thick disk or of the halo.)

Spitzer and Schwarzschild worked out a theory based on solving the Fokker–Planck equation (which describes the diffusion in velocity space that results from random stellar encounters) and following the increase in the velocity dispersions of the stars. They concluded that if the existing interstellar material were organized into complexes of a million solar masses each, their theory could account for the velocity dispersions of old disk stars.

Since this simple beginning, however, the theory of increase of velocity dispersions has become much more complicated. For one thing, Spitzer and Schwarzschild had solved only the two-dimensional problem in the Galactic plane. This omits an important question. As we have seen, the ratio of dispersions in the U and V components is set by the differential rotation of the Galaxy. The W dispersion, however, is determined by just the process that we are discussing. For it, different theoretical treatments have given different results.

Another problem has been the time dependence of the increase in velocity dispersions. Again, different theoretical treatments have given different results; furthermore, there is not complete agreement on the observational data. For the time dependence, see a discussion by Wielen (1990), and for the theory see the discussion by Binney and Lacey (1988).

It is clear that we do not yet fully understand this intriguing problem.

THEORIES OF SPIRAL STRUCTURE

Although spiral galaxies are so common, and although it was realized many decades ago that the Milky Way is a spiral galaxy, the details of spiral structure have been perversely stubborn in falling into place. It was 1952 before the locations of spiral arms in the Milky Way began to emerge, and they are still seen only in a murky and confused way. But our understanding of the dynamics of spiral structure has evolved even more slowly.

As a start, we should examine the image of a well-resolved spiral galaxy. It is clear that the spiral arms are marked by H II regions and bright, young stars. In other words, the spiral arms are regions of star formation and of recently formed stars. Indeed, the old stars of a disk that has no gas (as, for instance, in an S0 galaxy) show no tendency toward spiral structure, which is characteristic only of star-forming galaxies.

This suggests strongly, then, that it is to the interstellar material and to the young stars that we should look for the origin of spiral structure. The nature of a theory is still not obvious, however, because these components differ from the older stars in several different ways.

(1) The gas plays an important role; is it possible that spiral structure is a phenomenon of gas dynamics rather than stellar dynamics?

(2) The young stars have not yet fully mixed into the general field; could spiral structure arise from the particular way in which their original spotty distribution changes to a smooth one? (Remember the way in which an association develops a trailing shape during much of the early dispersal time of its stars.)

(3) Interstellar gas and young stars both have lower velocity dispersions than old stars; does velocity dispersion play an important role in the nature of spiral structure? This question, in particular, is reminiscent of the Toomre criterion, and the way in which low velocities can lead to instability.

The Winding Dilemma

In addition to looking at the population makeup of spiral arms, we must also consider their kinematics. We cannot, of course, watch them move, but we can predict how they ought to move, and a serious difficulty immediately arises. If, in fact, we take a simple-minded view of the motion of spiral arms, it is hard to see how differential rotation can allow them to persist at all. We noted, when studying epicycles, that two regions whose distances from the Galactic center differ by ΔR have a relative drift rate of $-2A\Delta R$. Since $(2A)^{-1} \cong 35$ Myr, a material structure such as a spiral arm is subject to such a strong shear that the arm would become tightly wound in only the time of one galactic rotation or so. This difficulty is called the "winding dilemma." It shows that material arms cannot persist. If an arm consists of the same material throughout its life, it must soon wind up hopelessly. The prevalence of spiral structure in so many galaxies could then be explained only by postulating a mechanism for continually generating new arms. But then we would expect to find, in the same galaxy, arms at all stages of winding; nor would we see the well-observed correlation between galaxy type and the openness of the arms.

The Density Wave Theory

One way out of the winding dilemma is to give up the idea that a spiral arm always consists of the same material. It might rather be the locus of a phenomenon that moves through the material of the disc, keeping the same spiral shape as it moves. This is the basis of the *density wave* theory of Lin and Shu, which envisions a *quasistationary spiral structure* (an idea that goes back to Lindblad), in which a wave of compression moves through the stellar and interstellar material as a configuration whose shape stays fixed, in a framework that rotates with "pattern speed" Ω_p, which is less than the material speed of galactic rotation. Thus a particular point in the material sees density waves go by (in a direction opposite to galactic rotation, because the pattern speed is slower) at an angular rate $2(\Omega - \Omega_p)$, where 2 is the number of arms in a "grand design" spiral.

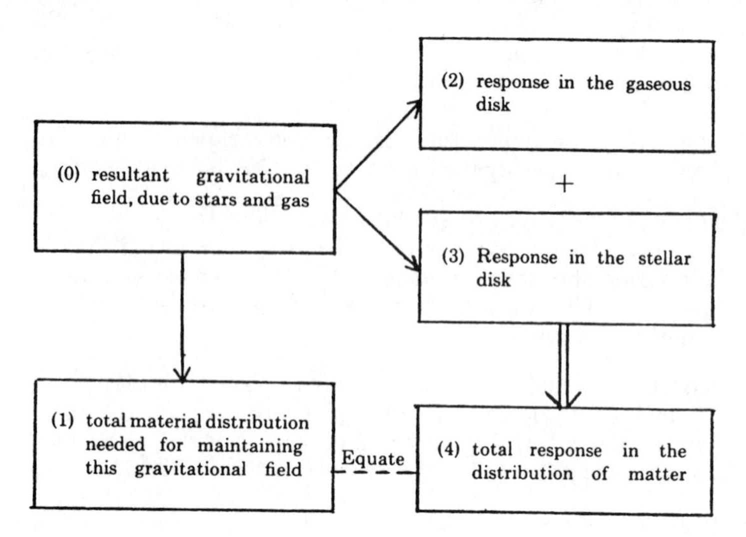

FIGURE 9 Logical outline of the relationships in the Lin–Shu density-wave theory. (From Lin 1966.)

Of particular importance to this linear theory are the regions where resonances occur, because there the theory blows up, and (through much more complicated, non-linear, calculations) there could be amplification, reflection, absorption, or even creation of waves. The most important of these resonances are *corotation* (rather far out, in most density-wave fits to a galaxy), where $\Omega = \Omega_p$ and the density wave remains stationary with respect to the material, and the *inner Lindblad resonance*, where $\Omega_p = \Omega - \kappa/2$. At the latter point, density waves pass through in synchronism with the local epicycle frequency κ.

The basic density-wave theory of Lin and Shu assumes small-amplitude sinusoids and fits them so that the density perturbations, the gravitational perturbations, and the motion perturbations are all mutually consistent. Lin (1966) describes the process with a logical diagram showing the interconnections that are needed if the theory is to be self-consistent (Fig. 9). Note that the quantities referred to are not the total gravitation or density but rather the increments that are associated with the density wave. Lin starts with the gravitation (labeled 0). It has two aspects. First, it corresponds, through Poisson's equation, to the density (1). Second, it induces density changes, which are in two parts. The gas response to the gravitational perturbation (2) is calculated, in a reasonably straightforward way, by perturbing the hydrodynamic equations of a fluid. The stellar response (3) is found from a perturbed form of the collisionless Boltzmann equation; it is a rather complicated calculation. The stellar response can be quite important, and its amount is sensitive to how close to the Toomre criterion the stars are. The final step is to insist that the total induced density (4) be equal to the density (1) that causes it.

The mathematical details of the density-wave theory are much too complicated to present here. Suffice it to say that they have been worked out in some detail by Lin, Shu, and others. (See, *e.g.*, Lin 1966.)

At this point it should be clearly understood what the basic linear density-wave theory contributes to an understanding of spiral structure and what it does not contribute. It successfully describes a set of waves that are dynamically self-consistent, in that they arouse density perturbations that are just right to provide the gravitational forces that are needed in order to cause those perturbations. The theory does not, however, predict the amplitude of the waves; in a linear theory such as this, the amplitude is arbitrary (as long as it is small enough to justify the approximations that are made). Nor does the theory predict that such waves must arise. The waves are also neutral waves, in the sense of a mathematical stability analysis; they neither grow nor decay. In short, the linear theory of density waves says neither that the waves must arise, nor how strong they should be, nor how they were generated, nor what can maintain them.

To attack these problems, it is necessary to work at the level of a non-linear theory. Lin and his associates have for many years sought modes that will grow spontaneously into "grand-design" spiral patterns (see, *e.g.*, Lin and Bertin 1985). Others have investigated mechanisms, independent of the density-wave theory, by which a more-local irregularity can grow and be sheared into a spiral form. The most interesting of these is "swing amplification" (see, *e.g.*, Toomre 1981). In so far as it is understood, this process is based on a combination of self-gravity, shearing, and the tendency for radial structures to grow (and therefore to be strongest when they are trailing).

Forms of Spiral Galaxies

The eventual truth might well draw on more than one process, in fact. After all, the galaxies themselves show different forms of spiral pattern (Elmegreen and Elmegreen 1982). The neat two-armed "grand-design" spirals are a minority; a more typical spiral has epidemic spirality but no clear, continuous arms. The name given to spirals of this type is "flocculent." The extreme of this tendency is exhibited by galaxies that are surrounded by almost-Medusa-like sets of arms. (See Figs. 10 and 11.) It has frequently been suggested (see, *e.g.*, Kormendy and Norman 1979) that grand-design spirals are produced by bars or tidal encounters, and that a galaxy left to itself will produce flocculent spiral structure.

In between, however, is a large class of galaxies that show a strong spirality, but no clear, continuous spiral arms. (See, for example, Fig. 12.) It may well be that the Milky Way is of this type.

Dynamical problems apart, however, the basic density-wave theory can at least be looked upon as giving a kinematic description of any quasistationary spiral structure. If arms are to exist without winding up, they will move in this way and will set up the patterns of density and velocity perturbations that are described by the density-wave theory. For plausible values of the density contrast between arms and interarm regions, the local streaming velocities in the gas are of the order of 5 to 10 km sec^{-1}. There should also be similar local perturbations in the velocities of low-velocity stars, depending on where they are in relation to a spiral arm. The stars of higher velocity should be affected less.

Since the Oort constants, particularly A, are calculated from stars of low velocity, their values might be perturbed by the streaming associated with a density wave, and perhaps do not correspond to the shape of the Galactic rotation curve in the way that they would in an axisymmetric galaxy—if, indeed, the Milky Way is a density-wave spiral.

Another important application of the density-wave theory has come from the fact that if the amplitude of a density wave is large enough, the velocities can be supersonic. (The "speed of sound" is rather close to the atomic speed in an ionized medium, or the cloud speed in a cool medium, where the clouds are the "particles" that carry the density information. A rough value is 10 km sec^{-1}.) It is widely believed that shocking by the density wave can set off star formation in the spiral arms.

CONCLUSION

This has been far too brief a review to do justice to the subject of Galactic dynamics. Again, for more detailed accounts the reader is referred to the sources cited in the introduction.

REFERENCES

Athanassoula, E., and Sellwood, J. A. 1986, *Monthly Notices Roy. Astron. Soc.* **221**, 213.

Bahcall, J. N. 1984, *Astrophys. J.* **276**, 169.

Binney, J., and Lacey, C. 1988, *Monthly Notices Roy. Astron. Soc.* **230**, 597.

Binney, J., and Tremaine, S. 1987, *Galactic Dynamics.* Princeton: Princeton Univ. Press.

Blaauw, A. 1952, *Bull. Astron. Inst. Netherlands,* **11**, 414 (no. 433).

Blaauw, A., and Schmidt, M. (eds.) 1965, *Galactic Structure.* Chicago: Univ. of Chicago Press.

Bosma, A. 1981, *Astron. J.* **86**, 1825.

Burton, W. B. 1988, in Verschuur and Kellerman, p. 295.

Chandrasekhar, S. 1960, *Principles of Stellar Dynamics.* New York: Dover.

Contopoulos, G. 1958, *Stockholm Obs. Annals,* **20**, no. 5.

Djorgovski, S., and Sosin, C. 1989, *Astrophys. J. (Letters)* **341**, L13.

Eggen, O. J. 1965, in Blaauw and Schmidt, p. 111.

Elmegreen, D. M., and Elmegreen, B. G. 1982, *Monthly Notices Roy. Astron. Soc.* **201**, 1021.

Fall, S. M., and Lynden-Bell, D., eds. 1981, *The Structure and Evolution of Normal Galaxies.* Cambridge, Eng.: Cambridge Univ. Press.

Gilmore, G., King, I. R., and van der Kruit, P. C. 1990, *The Milky Way As a Galaxy.* Mill Valley, CA: University Science Books.

Gould, A. 1989a, *Astrophys. J.* **341**, 748.

———. 1989b, in Philip and Lu, p. 19.

Hénon, M., and Heiles, C. 1964, *Astron. J.* **69**, 73.

Hunter, C., and Toomre, A. 1969, *Astrophys. J.* **155**, 747.

Jeans, J. H. 1929. *Astronomy and Cosmogony.* Cambridge, Eng.: Cambridge Univ. Press.

King, I. R. 1993, *Introduction to Stellar Dynamics*, in preparation.

Kuijken, K., and Gilmore, G. 1989a, *Monthly Notices Roy. Astron. Soc.* , **239**, 571.

Kuijken, K., and Gilmore, G. 1989b, *Monthly Notices Roy. Astron. Soc.* , **239**, 605.

Kuijken, K., and Gilmore, G. 1989c, *Monthly Notices Roy. Astron. Soc.* , **239**, 651.

Kuijken, K. 1991a, *Astrophys. J.* **372**, 125.

Kuijken, K. 1991b, *Astrophys. J.* **376**, 467.

Lewis, J. R., and Freeman, K. C. 1989, *Astron. J.* **97**, 139.

Kormendy, J., and Norman, C. A. 1979, *Astrophys. J.* **233**, 539.

Lin, C. C. 1966, *Lectures in Applied Mathematics* **9**, pt. 2, 66.

Lin, C. C., and Bertin, G. 1985, in van Woerden *et al.*, p. 513.

McCuskey, S. W. 1965, in Blaauw and Schmidt, p. 1.

Nordström, B., and Andersen, J. 1989, in Philip and Lu, p. 133.

Ollongren, A. 1965, in Oort 1965, pp. 501–509.

Oort, J. H. 1932, *Bull. Astron. Inst. Netherlands*, **6**, 249 (no. 238).

———. 1960, *Bull. Astron. Inst. Netherlands*, **15**, 45 (no. 494).

———. 1965, in Blaauw and Schmidt, p. 455.

Ostriker, J. P., and Peebles, P. J. E. 1973, *Astrophys. J.* **186**, 467.

Philip, A. G. D., and Lu, P. (eds.). 1989, *The Gravitational Force Perpendicular to the Galactic Plane.* Schenectady: L. Davis Press.

Sandage, A. 1957, *Astrophys. J.* **125**, 435.

Sparke, L. S., and Casertano, S. 1988, *Monthly Notices Roy. Astron. Soc.* **234**, 873.

Spitzer, L., and Schwarzschild, M. 1951, *Astrophys. J.* **114**, 385.

———. 1951, *Astrophys. J.* **118**, 106.

Statler, T. S. 1989a, *Astrophys. J.* **344**, 217.

———. 1989b, in Philip and Lu, p. 133.

Toomre, A. 1964, *Astrophys. J.* **139**, 1217.

———. 1981, in Fall and Lynden-Bell, p. 111.

Verschuur, G., and Kellerman, K., eds. 1988, *Galactic and Extragalactic Radio Astronomy.* New York: Springer.

Wielen, R. ed. 1990, *Dynamics and Interactions of Galaxies.* Heidelberg: Springer.

Wielen, R., and Fuchs, B. 1990, in Wielen 1990, p. 318.

Woerden, H. van, Allen, R. J., and Burton, W. B., eds. 1985. *The Milky Way Galaxy.* I. A. U. Symposium no 106. Dordrecht: Reidel.

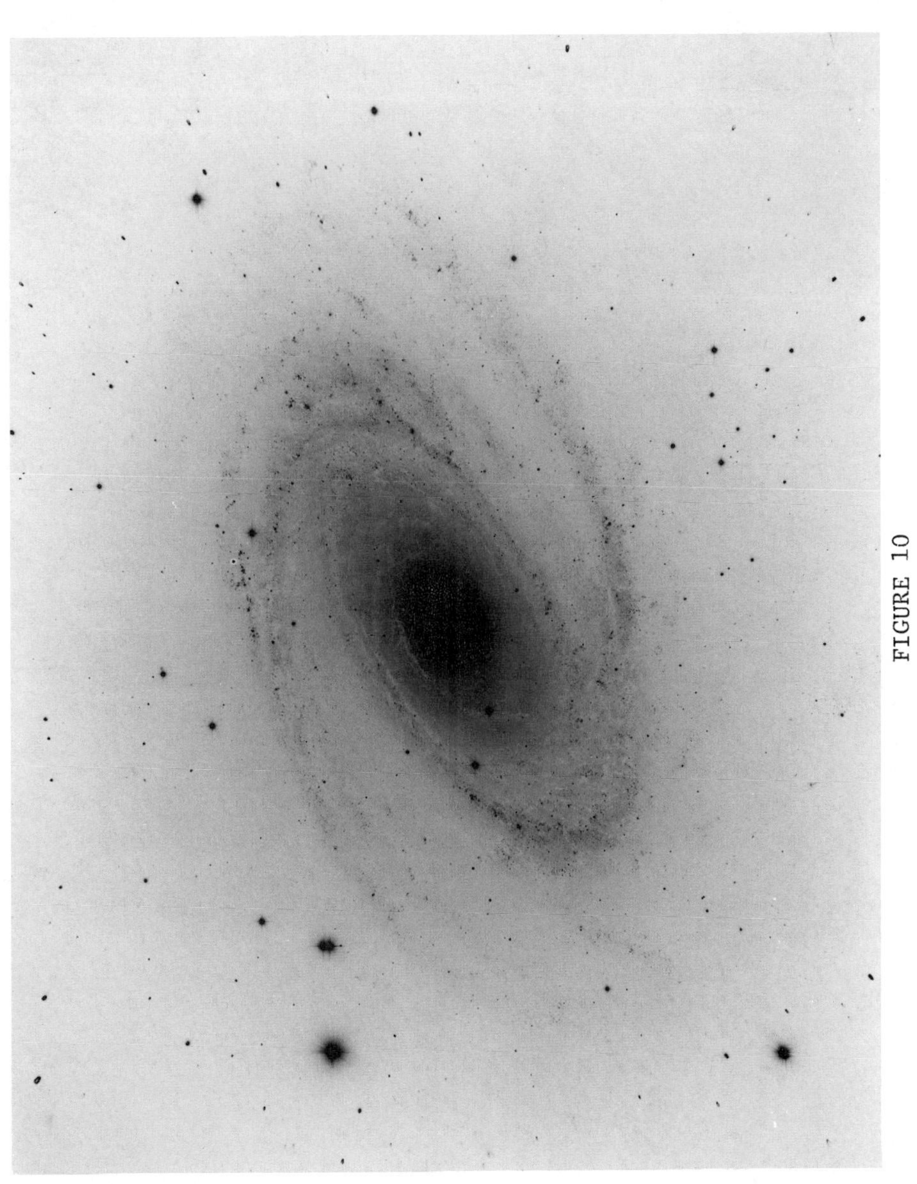

FIGURE 10

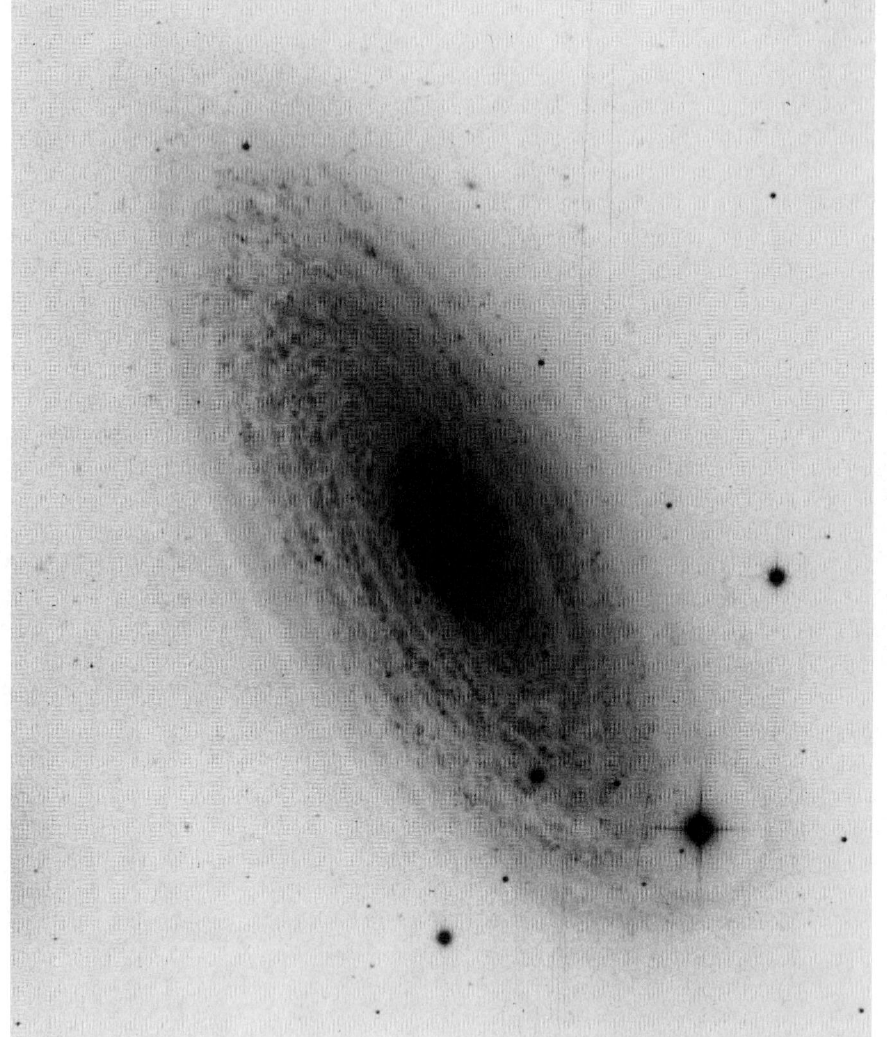

FIGURE 11

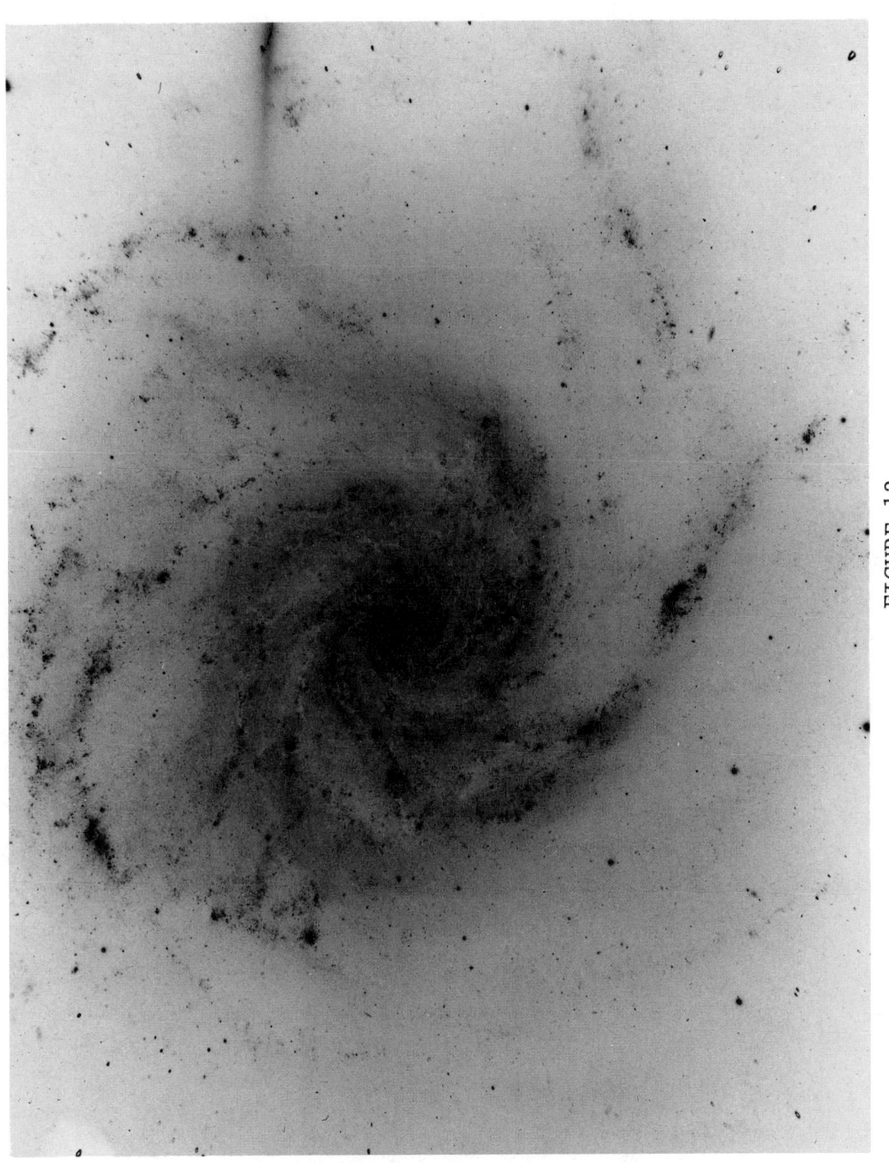

FIGURE 12

QUESTIONS AND PROBLEMS

1. A "Fermi Problem" is a mid-twentieth century American art form attributed to the physicist Enrico Fermi (1901-1954). The idea is to answer a question about the world, usually numerical in content, using no reference books and in general using minimal input information. The most famous Fermi Problem embodied in his casual 1943 remark "Where Are They?", has led to many well-known books. A purer example (which seems at first sight to be more mundane) is, "How many piano tuners are listed in the Chicago telephone directory?" -- the answer to be estimated without reference to any book or other informant whatsoever. The answers to Fermi Problems may vary widely in significance.

The following questions about the Galaxy are not really Fermi Problems, but they can be regarded in the same spirit: try to make rough numerical estimates using almost no reference material. (Some numbers such as the Solar mass and luminosity should be available in your head.) Some parts do require reference input, but these can be regarded as an exercise in locating crucial astrophysical data in a library or elsewhere.

The Problem: PRESSURES and ENERGY DENSITIES are usually comparable in some sense; they have the same dimensions and units. Good units for this problem are "eV per cubic centimeter" (1 eV = 1.6e-12 erg, roughly the energy of a 1-micron-wavelength photon). Evaluate these as best you can.

a) What is a typical gas pressure in the Galactic Disk? (For cold components this should be separated into turbulent and purely thermal contributions.)

b) What is the energy density in the magnetic field in the Galactic Disk?

c) What is the energy density of starlight in the Galactic Disk?

d) What is the energy density of cosmic ray particles in the Galactic Disk?

e) What is the energy density of the cosmic 3-K background?

f) What energy densities matter in the Galaxy, and how?

2. Investigate the uncertainty in the age of the globular cluster M92. Examine at least the following factors:

Overall Photometric precision
Interstellar Reddening
Composition (Metallicity)
Oxygen abundance
Helium abundance
The distance scale
Theoretical isochrones
Effective temperature - color transformations

Consider the probable uncertainty in each parameter and its importance in affecting the age of the cluster. You may find Stetson and Harris (1988, A.J., 96, 909) to be a useful reference.

3. The globular cluster NGC 5694 is located 26±5 kpc from the galactic center at $\ell = 332°$, $b = 30°$. Its radial velocity is -180±20 km sec^{-1}. Assume that the galaxy is spherical with a flat rotation curve out to some maximum distance R_*. Outward from R_* the rotation curve is Keplerian. Within R_*, the circular speed is $V_c = 220$ km sec^{-1}, independently of the distance from the galactic center.

a) Assuming that the cluster is bound to the galaxy, calculate a lower limit to R_*.

b) Show that the total density of matter in the galactic halo, $p\mathcal{H}$ follows a law of the form, p^{-2} under the above assumptions.

c) Calculate the total mass of the galaxy at R_*.

4. What is Malmquist Bias and how does one go about identifying it and then correcting for it in any data set?

5. Determine the distance R_0 to the galactic center, assuming that the motions of bulge stars are isotropic. In Baade's window, the radial velocity dispersion is

$$\sigma_r = 113 \pm 6 \text{ km/s}$$

From recent proper motion surveys

$$\sigma_\mu = 0.32 \pm 0.01 \text{ arcsec/century}$$

$$\sigma_\mu = 0.28 \pm 0.01 \text{ arcsec/century}$$

Calculate the error in R_o also.

6. Consider the galaxy as being composed of a spherical bulge with $\rho(R) \propto R^{-3}$ (cylindrical coordinates) and a constant surface density, very thin disk. Calculate

a) the height Z_c along the symmetry axis, and
b) the distance R_c along the plane

at which the gravitational potential from the disk exceeds that from the bulge. Since this is to be a rough calculation, assume that in the disk the potential at a given radius R depends only on the mass within R. Use this parameter:

DISK: Surface density is 200 $M_\odot/pc^2$

BULGE: Volume density is 10 $M_\odot/pc^{-3}$ at R = 500 pc

7. Is the bulge really a collisionless system? Suppose the stellar density deep in the bulge is 40 $M_\odot$ pc^{-3} and that each star has a mass of 1 $M_\odot$. Assume stars have a random velocity of 100 km s^{-1} and that they "collide" (i.e. have a change in their orbits) if they pass within 100 A.U. of one another. (Hint: consider how much volume per unit time is swept up by an object of size 100 A.U. moving at 100 km s^{-1}.)

8. a) Explain the intuitive physics that underlies the determination of the local mass density ρ_0 and the local surface density σ_0 from observations of stellar motions and density gradients perpendicular to the Galactic plane.

b) What are the practical difficulties in these determinations, and why are present-day results so much less than satisfactory?

9. What arguments make it clear that spiral structure must involve patterns that move through the stellar disk? What are the advantages of the linear density wave theory, and in what ways is it inadequate?

10. Discuss why there is an effective absence of an age-metallicity relation for disk stars with ages between 10 Gyr and 1 Gyr.

Index